Springer Theses

Recognizing Outstanding Ph.D. Research

Aims and Scope

The series "Springer Theses" brings together a selection of the very best Ph.D. theses from around the world and across the physical sciences. Nominated and endorsed by two recognized specialists, each published volume has been selected for its scientific excellence and the high impact of its contents for the pertinent field of research. For greater accessibility to non-specialists, the published versions include an extended introduction, as well as a foreword by the student's supervisor explaining the special relevance of the work for the field. As a whole, the series will provide a valuable resource both for newcomers to the research fields described, and for other scientists seeking detailed background information on special questions. Finally, it provides an accredited documentation of the valuable contributions made by today's younger generation of scientists.

Theses are accepted into the series by invited nomination only and must fulfill all of the following criteria

- They must be written in good English.
- The topic should fall within the confines of Chemistry, Physics, Earth Sciences, Engineering and related interdisciplinary fields such as Materials, Nanoscience, Chemical Engineering, Complex Systems and Biophysics.
- The work reported in the thesis must represent a significant scientific advance.
- If the thesis includes previously published material, permission to reproduce this must be gained from the respective copyright holder.
- They must have been examined and passed during the 12 months prior to nomination.
- Each thesis should include a foreword by the supervisor outlining the significance of its content.
- The theses should have a clearly defined structure including an introduction accessible to scientists not expert in that particular field.

More information about this series at http://www.springer.com/series/8790

Wenfei Jin

Admixture Dynamics, Natural Selection and Diseases in Admixed Populations

Doctoral Thesis accepted by
the University of Chinese Academy of Sciences, China

Springer

Author
Dr. Wenfei Jin
CAS-MPG Partner Institute
for Computational Biology (PICB),
Shanghai Institutes for Biological
Sciences, Chinese Academy
of Sciences
Shanghai
China

Supervisors
Prof. Li Jin
PICB, Shanghai Institutes for Biological
Sciences, Chinese Academy of Sciences
Fudan University
Shanghai
China

Prof. Shuhua Xu
PICB, Shanghai Institutes for Biological
Sciences, Chinese Academy of Sciences
Shanghai
China

ISSN 2190-5053 ISSN 2190-5061 (electronic)
Springer Theses
ISBN 978-94-017-7406-2 ISBN 978-94-017-7408-6 (eBook)
DOI 10.1007/978-94-017-7408-6

Library of Congress Control Number: 2015950039

Springer Dordrecht Heidelberg New York London

Printed on acid-free paper

Springer Science+Business Media B.V. Dordrecht is part of Springer Science+Business Media (www.springer.com)

Parts of this thesis have been published in the following journal articles:

[1] ***Wenfei Jin***, *Shuhua Xu, Haifeng Wang, Yongguo Yu, Yiping Shen, Bailin Wu, Li Jin.* 2012. Genome-Wide Detection of Natural Selection in African Americans Pre-and Post-Admixture. *Genome Res.* 22: 519–527.

[2] ***Wenfei Jin***, *Pengfei Qin, Haiyi Lou, Li Jin, Shuhua Xu.* 2012. A Systematic Characterization of Genes Underlying Both Complex and Mendelian Diseases. *Hum Mol Genet.* 21: 1611–1624.

[3] ***Wenfei Jin***, *Sijia Wang, Haifeng Wang, Li Jin, Shuhua Xu.* 2012. Exploring Population Admixture Dynamics via Empirical and Simulated Genome-Wide Distribution of Ancestral Chromosomal Segments. *Am J Hum Genet.* 91: 849–862

[4] ***Wenfei Jin#***, *Ran Li#, Ying Zhou, Shuhua Xu.* 2014. Distribution of ancestral chromosomal segments in admixed genomes and its implications for inferring population history and admixture mapping. *Euro J Hum Genet.* 22, 930–937. (# equal contribution)

[5] *Shuhua Xu,* ***Wenfei Jin***, *Li Jin.* 2009. Haplotype Sharing Analysis Showing Uyghurs Are Unlikely Genetic Donors. *Mol Biol Evol.* 26: 2197–2206.

[6] *Shuhua Xu,* ***Wenfei Jin***. 2012. Population Genetics in the Genomic Era. Studies in Population Genetics, M. Carmen Fusté (Ed.). ISBN 978-953-51-0588-6. Intech. UK

Supervisor's Foreword I

Modern humans have successively colonized almost every corner of the earth since the out of Africa about 100,000 years ago. The geographic isolation and adaption to local environment in the past tens of thousands of years have shaped our human species, leading humans in different regions/environments having developed into distinct populations with genetic and phenotypic differences. Following the age of discovery, people from different continents could meet and intermarry in the New World. Admixed populations, such as African Americans and Latinos, have come into being and compose a significant fraction of the humans in the New World. These immigrants and their descents lived in the New World with completely different environment compared with where their ancestry evolved, which may implicate they are subjected to strong natural selection. Extending from this interesting question, the dissertation vividly showed the distribution of length of ancestral chromosomal segments in admixed population, population admixture dynamics, signatures of natural selection, as well as the natural selection on genes underlying different disease categories. In summary, the dissertation significantly increased our knowledge about the admixed populations, especially African Americans. I hope that the study will allow many people to understand, and eventually contribute to, the theory of population admixture and natural selection in the modern humans.

Fudan, Shanghai
August 2015

Prof. Li Jin

Supervisor's Foreword II

The immigrants and their descents lived in the New World with quite different environment compared with what their ancestors have experienced; it would not be surprising if natural selection have occurred in their admixed descendants. In addition, it is well known that the African ancestors of African Americans experienced a bottleneck (high mortality) during the Middle Passage and the first several years after they arrived in the America. If the selection forces acting on the admixed population were strong enough and consistent, it is possible to detect the footprints of selection in the recently admixed populations.

However, detecting natural selection in admixed populations requires accurate inference of local ancestry and admixture history. In Wenfei's work for his Ph.D. thesis, he realized that the distribution of length of ancestral of chromosomal segments (LACS) in the admixed population is useful for inferring the population admixture dynamics. By comparing the distribution of LACS in admixed population with various hypothesized scenarios, he was able to decide the scenarios with distribution of LACS best fit the observed distribution. Wenfei went beyond detecting the post-admixture selection signatures by examining the ancestry deviation in the admixed population. He has constructed an artificial African population using the inferred African ancestral chromosomal segments and compared it with indigenous African populations, with which he was able to detect the natural selection since the African ancestors of African American left for the New World. His analyses showed that the genetic variants protecting against malaria in the artificial African population were lower in frequency than that in the indigenous African population, consisting the relaxation of selection pressure of malaria in African American population. Especially, many genes showing signatures of natural selection are associated with African American specific high-risk diseases such as prostate cancer and hypertension, which led Wenfei to do a systematic study of genes associated with diseases. Analyses of human polymorphic data, species divergence data, copy number variations data, gene expression data, and interaction network showed that the genes underlying both monogenic diseases and complex diseases are subject to both strong positive and purifying selections, which probably have occurred at different times in human history; while genes underlying only monogenic diseases and only complex diseases were subjected to strong purifying selection and positive selection, respectively.

The dissertation revealed the distribution of LACS in admixed population, population admixture and dynamics, signatures of natural selection, as well as the natural selection on genes associated with different diseases. I expect that the study presented here will significantly advance our understanding of the processes and mechanism of population admixture and natural selection in the modern humans.

Shanghai
December 2014

Prof. Shuhua Xu

Abstract

Admixed populations, such as African Americans and Latinos, have come into being and compose a significant fraction of humans in the New World. We found that the distribution of lengths of ancestral chromosomal segments (LACS) in the admixed population could provide sufficient information for inferring the population admixture dynamics. In order to understand the distribution of LACS, we introduced a theoretical framework on the distribution of LACS in hybrid isolation model and gradual admixture model. Although the distribution of LACS in gradual admixture model differed from that in hybrid isolation model, we showed that the mean LACS in gradual admixture model was approximately half that of hybrid isolation model, when both admixture proportion and admixture time in the two models were identical. By comparing the distribution of LACS in admixed population with those of various hypothesized scenarios, we found the scenario with distribution of LACS to best fit the observed distribution. Based on 491,526 autosomal SNPs genotyped in thousands of African Americans, we found that a continuous gene flow model, in which African Americans continuously received gene flow from European parental population over about 14 generations, best explained the admixture dynamics of African Americans among several putative scenarios. Interestingly, we observed that some African Americans had much more European ancestry than that of the simulated, indicating substructures of local ancestries in African Americans that could be resulted from individuals from some particular lineages having continuously intermarried with people of European ancestry. On the contrary, the admixture dynamics of Mexicans was more likely to be explained by a gradual admixture model in which Mexicans continuously received gene flow from both European and Amerindian parental populations for about 24 generations.

It is particularly meaningful to investigate the signatures of natural selection in African Americans as their ancestry experienced high mortality. The several genomic regions showing excess of African or European ancestry could be the footprints of natural selection in the post-admixture era. In order to detect the natural selection since the African or European ancestry left for the New World, we have reconstructed an artificial African population using the inferred African ancestral chromosomal segments and compared it with indigenous African populations. Interestingly, many selection-candidate genes identified by the latter approach were associated with African American specific high-risk diseases such as prostate cancer and hypertension, suggesting these disease-related genes might have played an important role in African Americans adapting to the new environment. *CD36* and *HBB*, whose mutations confer a degree of protection against malaria, were also located in those highly differentiated regions between the artificial African population and the indigenous African population. Further analysis showed that frequencies of alleles protecting against malaria were lower in artificial African population than that in indigenous African population, which consists of the relaxed selection pressure of malaria in the U.S.

Because we found many genes associated with disease have been subjected to natural selection in African Americans, we performed a systematic analysis on genes associated with diseases. We found that the number of genes underlying both monogenic and complex disease (MC genes) is significantly overrepresented compared with that by random. Analyses of human polymorphic data, species divergence data, copy number variations (CNV) data, gene expression data and interaction network showed that MC genes are subjected to both strong positive and strong purifying selections, simultaneously; while genes underlying only monogenic diseases or only complex diseases were subjected to strong purifying selection and strong positive selection, respectively. For examples, the genes only underlying monogenic diseases are significantly impoverished around CNV regions because of purifying selection; the genes only underlying complex diseases are significantly enriched around CNV regions because of positive selection. The MC genes are neither enriched nor impoverished around the CNV regions because they are subjected to both strong positive selection and purifying selection.

In summary, this work not only contributes greatly to our knowledge on recent human population admixture, nature selection in African Americans and characteristic of genes undying diseases but the approach in this study also could apply to other admixed populations, e.g., the theoretical the distribution of LACS in admixed population facilitates the inference of population admixture dynamics.

Keywords Admixed population · Population admixture · Ancestral chromosomal segment · Admixture model · Hybrid isolation (HI) model · Gradual admixture (GA) model · Continuous gene flow (CGF) model · Admixture mapping · African Americans · Monogenic disease · Complex disease · Disease gene · Evolution · Copy number variation (CNV) · Tissue specificity · Ingenuity pathway analysis (IPA)

Acknowledgments

It is a great honor to be one of the first Ph.D. students enrolled and graduated from PICB. I would like to express my sincere appreciation to those people who made this thesis possible with their support and assistance.

First and foremost, I want to thank my supervisors Prof. Felix Li Jin and Dr. Shuhua Xu. With their enthusiasm, inspiration, and endeavor, they made population genetics fun for me. I still vividly remember many of the meetings and conversations with them. I appreciate both of them for their guidance and suggestions, listening to my boring reports and reading pages of drafts. Particularly, I thank Felix for his ideas, patience, kindness, and academic experience, which are invaluable. I still remember many interesting stories about his experiences in research and science, which are enlightening. Dr. Xu is the guide who led me to the kingdom of science and made my first publication come true. I also appreciate him for paying so much attention to me and designing my initial Ph.D. project. I am also thankful for the excellent examples both supervisors have set themselves as successful population geneticists. Finally, I really appreciate their patience for revising my manuscripts, which could not have been published without their help and hard work.

The members of the Computational Genomics Group and Population Genomics group have contributed immensely to my personal and professional time. Both the groups always provide source of good advice and collaboration, as well as friendship. First, I am grateful for people who have given me constructive suggestions including Zhengwen Jiang, Yungang He, Shi Yan, Wei Wang, Erli Wang, and Ran Li. I also would like to acknowledge other members in the group including Jing Pu, Ling Yang, Haiyi Lou, Pengfei Qin, Hongyang Xu, Dongsheng Lu, Minxian Wang, Meng Shi, Ying Zhou, and Ruiqing Fu. The encouragements from my friends have been indispensable, and I would like particularly to acknowledge: Guofeng Meng, Yuling Liu, Chunxuan Shao, Chaofeng Wang, Kai Weng, Zhongshan Li, Ningyi Shao and Jing Guo. I also acknowledge Ms. Lisa Li, who has provided much help in both my life and my study so that I can concentrate on my study and research.

I also appreciate my brothers, sisters, aunts, and uncles for their support. Especially, I thank my second elder brother for his strong support during the last 10 years and hope he has a more bright future.

Last but most important, I want to appreciate my father, who has devoted so much to me that I cannot overstate my gratitude to him. I feel proud that I have such a great father.

Contents

Abbreviations

AAF	Artificial African population
CNV	Copy number variation
GAD	Genetic association database
GWAS	Genome-wide association studies
HapMap	International haplotype map project
HGDP	Human genome diversity project
hOMIM	Hand-curated OMIM database
HPRD	Human protein reference database
iHS	Integrated haplotype score
IPA	Ingenuity pathway analysis
MC	Monogenic and complex diseases
OMIM	Online mendelian inheritance in man database
PCA	Principle component analysis
rAfA	Reconstituted African Americans
SEM	Standard error of the mean
SNP	Single nucleotide polymorphism

Chapter 1
Introduction

Abstract Modern humans have successively colonized almost every corner of the earth since the out of African about 100,000 years ago. However, there are various distinct environments on the earth, varying in light, temperature, moisture, and so on. The geographic isolation and adaption to local environment in the past tens of thousands of years have shaped our human species, leading human beings in different regions/environments having developed into distinct populations with genetic and phenotypic difference. Following the Age of Discovery, admixed populations, such as African American and Latinos, came into being and composed a significant fraction of the human beings in the New World. These admixed populations provided exciting opportunities to identify the genetic variation underlying complex diseases. With the advance of the genomic genotyping and sequencing technology, the statistical powers for identifying genetic variants underlying human disease have significantly improved, using either associate study or admixture mapping.

Keywords Human evolution · Population structure · Population admixture · Gene mapping · Disease susceptibility variants

1.1 Origin of Modern Human

Our modern humans (Homo sapiens) are the only extant species of the *genus* Homo. Therefore, human origins include both the emergences of genus *Homo* and anatomically modern human. It is known that genus *Homo* diverged from Australopithecines 2–3 million years ago in Africa [1, 2]. Molecular evidences suggested that the closest surviving cousin of human is chimpanzee, whose ancestry split from that of human 4–7 million years ago [3]. *Homo erectus,* with brain size increasing gradually, was believed to be the first Homo species using fire and complex tools. Furthermore, *Homo erectus* left Africa and spread to Asia and

W. Jin, *Admixture Dynamics, Natural Selection and Diseases in Admixed Populations*, Springer Theses, DOI 10.1007/978-94-017-7408-6_1

Europe about 1.3–1.8 million years ago [4]. They gradually evolved to archaic humans such as Heidelberg Man and Neanderthals at different regions and periods.

Modern humans evolved from archaic humans and first appeared in Africa about 200,000 years ago. Anatomically modern humans are distinguished from archaic humans by having a thin skull, decreased brow ridges, and a prominent chin. However, scientists did not reach a consensus on whether modern humans originated solely in Africa until the beginning of twenty-first century [5, 6]. There are always two competing hypotheses against each other on the origin of modern human: multiregional evolution and recent African origin. The former holds that local archaic humans in each region evolved to modern humans; while the latter proposes that all modern humans originated in Africa and later dispersed into the other parts of the world. The pioneering work on mtDNA variations by Cann et al. [7] in the 1980s is a milestone in human evolution study due to its implications for the understanding of modern human origin and dispersal. Recent African origin hypothesis has been widely accepted and has become the mainstream in the twenty-first century because analyses of mtDNA and Y chromosome both supported this hypothesis [5, 7, 8]. According to recent African origin hypothesis: archaic humans evolved to anatomically modern humans solely in Africa between 200,000 and 150,000 years ago. Then a branch of modern humans left Africa between 125,000 and 60,000 years ago. They co-existed with the other archaic humans such as Neanderthals and *Homo erectus* in Eurasia for a long time and finally replaced these local archaic humans.

However, the mtDNA or Y chromosome only represents a locus in the genome and has suffered from serious genetic drift, the analyses using genome-wide data have greatly enriched our knowledge of modern human origins and migration. In recent several years, the genomes of several extinct archaic humans have been sequenced [9, 10], although there is no consensus on whether these archaic humans are subspecies of modern humans or separate human species. The first sequenced archaic human was Neanderthal, the closest evolutionary cousin of the present anatomically modern human. The Neanderthal used to live in large parts of Europe and Western Asia before extinction 30,000 years ago. Analysis of the Neanderthal genome essentially supported the recent African origin of modern human. However, Neanderthals shared more genetic variants with present modern humans in Eurasia than that in Sub-Saharan Africa, which suggested that gene flow from Neanderthals into the ancestry of non-African occurred before the divergence of Eurasian groups [9]. It is estimated that 1–4 % of the genomes of individuals in present Eurasia is derived from Neanderthal. Another sequenced archaic human was referred to as Denisovans, who shared a common origin with Neanderthals, and the bone was found in Denisova Cave in southern Siberia [10]. Denisovans were not involved in the putative gene flow from Neanderthals to Eurasians; however, data analysis suggested that it contributed 4–6 % of its genetic material to present Melanesians. Later study demonstrated that gene flow from Denisovans has contributed to the gene pool of New Guineans, Australians, and Mamanwa, but not mainland East Asians, western Indonesians, Jehai, and Onge [11].

Sequencing of an aboriginal Australian genome showed that they are descendants of an early human dispersal into eastern Asia, possibly 62,000–75,000 years ago. This dispersal was separate from the one that gave rise to modern Asians 25,000–38,000 years ago [12]. Further analyses showed that gene flow between populations of the two dispersal waves was prior to the divergence of Native Americans from modern Asian ancestors. These results suggest present-day aboriginal Australians descended from the earliest humans to occupy Australia, likely representing one of the oldest continuous populations outside Africa.

In summary, recent studies based on whole genomic sequence data essentially support that modern humans solely originated in Africa about 200,000 years ago. However, gene flows from some archaic humans contributed to the gene pool of modern non-Africans, although the contributions are only a few percentages. There are at least two waves of migration out of Africa: the one approximately 62,000–125,000 years ago gave rise to the aboriginal Australian, the other about 25,000–38,000 years ago developed the modern Eurasian and American.

1.2 Human Populations and Population Structure

An ideal population is a single entity in which any pair of individuals from different sex could randomly mate and reproduce. However, human population certainly is not randomly mating because many resistance forces, such as geographic barriers and society segregation, significantly hinder the individual random mating, which lead to population structure. Elucidating human population structure is one of the main goals of population genetics study because analyses of population structure not only increase our knowledge of population relationships and population history, but could also reduce false positive rates caused by population stratification in association studies. The traditional methods to study population structure could be classified into phylogeography and summary statistics [13]. The phylogeography focuses on the building of gene tree at population level, which is similar to the phylogenetic tree. It is especially powerful for analyzing mtDNA and Y chromosome that inherited as single "linkage block" without recombination. However, mtDNA and Y chromosome only account for a small fraction of the total genome and suffered from strong genetic drift. Summary statistics are the estimators describing the mathematical models of the common ancestry of DNA sequence from multiple loci. The autosomes account for most of the human genetic materials and are composed by thousands of millions of loci that are relatively independent, which complement each other and together provided more complete information about the population history. For example, the decrease of genetic and haplotype diversity from East Africa to other parts of the world, and the decreased slopes of ancestral allele frequencies showed the route of the out of African [14, 15]. Since the completion of the human genome project (HGP), the deluge of genetic data has driven the development of new statistic for exploring the human population history.

The human population structure potentially is one of the most complex ones because humans with a population size exceeded 7 billion and colonized almost every corner of the earth. The worldwide population samples were also essential for studying the global pattern of human population structure. Promoted by Cavalli-Sforza and his colleague in the 1990s, the human genome diversity project (HGDP) aimed to explore human diversity and population structure by looking at genomes from numerous indigenous populations across the earth. In order to offer the convenient and standard resources for laboratories working on human genetic variation, this project was designed to provide a shared set of DNA samples through immortalized lymphoblastoid cells [16]. Although it did not fulfill the initial goals because of controversies, HGDP already provided more than 1,000 samples in 53 indigenous populations from the world [17]. Recently, the genetic data from the international haplotype map project (HapMap) and the ongoing 1000 genome project have greatly facilitated the population genetic structure study.

Traditionally, most studies on human variation sampled different ethnics (populations) according to culture or geographic locations, and inferred population relationships based on these predefined populations, which may not reflect the true genetic relationships. However, the clustering methods such as STRUCTURE [18, 19], without using prior information about the origins of individuals, can directly infer the population genetic structure. STRUCTURE implements a model-based clustering method that integrated Markov chain Monte Carlo (MCMC) to infer population structure with multilocus genotypic data. It needs a long time to obtain the converged results and is computationally intensive. Rosenberg et al. [20] applied STRUCTURE on 1,056 individuals from 52 populations (HGDP panel) genotyping at 377 autosomal microsatellite loci. It is found that individuals from the same predefined populations always shared similar membership coefficient in inferred clusters. When the number of clusters was set to two, the clusters anchored by Africa and America, which are the two groups with the largest genetic distance. Increasing the number of clusters resulted in a previously single cluster separated into two. When the number of clusters was set to five, the five genetic clusters correspond to five geographic regions (Africa, Eurasia including Europe and Central/South Asia, East Asia, America, and Oceania) very well [20].

In order to take advantage of the genome-wide high-density data, many computationally efficient software and algorithms adapting to linkage disequilibrium have been developed [21, 22], which provide more detailed information about population structure [15, 23]. Using FRAPPE, the seven genetic clusters can correlate well with the seven major geographic regions, namely African, Middle East, Europe, Central/South Asia, East Asia, America, and Oceania [15]. Since individuals belonging to the same predefined populations in the same dataset show similar ancestry proportions, it is much useful to evaluate the genetic relationships among these populations. When a rooted tree was constructed using maximum likelihood method based on F_{ST}, the sub-Saharan Africans are located closest to the root of the tree, outward branches correspond sequentially to populations from North Africa, the Middle East, Europe, South/Central Asia, Oceania, America, and East Asia [15]. The pattern of branches consists of the orders of human expansion and migration route. Haplotypes are more likely to be geographically region-specific than individual SNPs [23]. The analysis based on the haplotype showed that human populations experienced series of founder effects, in which non-African population forms a sequential chain of colonies, which consists of that population far from African tending to have a smaller effective population size.

Multivariate techniques, such as principal component analysis (PCA), have been used to condense information of numerous loci into a few synthetic variables, which are especially powerful when analyzing the genome-wide high-density data. PCA has been introduced to population genetics by Cavalli-Sforza et al. [24] in 1978. The original PCA used populations as sample units, and population allele frequencies were analyzed to project the populations on the graph, which was an alternative method of phylogenetic trees. In 2006, Patterson et al. [25] implemented PCA on individual genotypic data and projected individuals on the graph, which made PCA much powerful tool to detect population structure and caught the scientist's interests. Reich et al. [26] suggested that PCA was very useful for population genetics and highlighted three applications: detecting population substructure, correcting for stratification in association studies, and making qualified inferences about human history.

By superimposing PCs on the geographic location of the studied populations, the synthetic maps on which are shown obvious gradients of variation across different regions may suggest historical population migrations. For example, the first PC map based on European populations showed a southeast-to-northwest cline and was interpreted as the reflection of Neolithic farming spread from the Levant to Europe about 6,000–9,000 years ago [24]. And the hypothesis about the expansion of Neolithic farming has been supported by many genetic and archeological data [26, 27]. However, according to the study by Novembre and Stephens [28], PCs correlating with geography do not necessarily reflect major population migrations but isolation by distance, in which gene exchanges are only among neighboring populations. For example, based on the dataset of 3,000 individuals genotyped at over half a million SNPs, Novembre et al. [29] found that the inferred principal components essentially reconstructed the geographic map of European. And they suggested that individual genome could be used to infer their geographic origin with surprising accuracy—often within a few hundred kilometers. Based on this situation, interpreting the PC maps is difficult and requires correlation of information from genetics with those from archeology and linguistics [26].

Although it is difficult to relate PCA to the underlying population events as it is non-parametric, a study by McVean [30] provided new insights into the relationship between PC projection and the underlying genealogical history. McVean showed that projection of individual samples onto the principal components could be obtained directly from considering the average coalescent times between pairs of haploid genomes. These results provide a framework for interpreting PCA projections in terms of underlying processes, including migration, geographical isolation, and admixture. McVean also demonstrated a link between PCA and Wright's F_{ST} [30].

In recent years, the methods and statistics that integrate linkage disequilibrium and haplotype information have been developed, which have promised to become the mainstream of population genetics in the future. These studies integrating haplotype information essentially have greatly improved our knowledge on the human history and population relationships. For example, the linkage disequilibrium information has been integrated into STRUCTURE as linkage model to estimate the ancestry along chromosome [18]. Based on haplotypes of their parental populations, the chromosomal segments of distinct ancestry in admixed population can be accurately inferred using population genetic model [31]. The lengths of migrant DNA tracts were used to infer the change of recent gene flow in different populations [32]. These approaches and strategies provide new insights into the application of haplotype information. Especially, based on a copying model adapted from Li and Stephens [33], the worldwide linkage disequilibrium pattern of human was used to infer human dispersal routes, which were consistent with "Out-of-Africa" bottleneck and a serial dilution of diversity during global colonization including a prominent East-Asian bottleneck [34]. The study also found some new points on the human history such as the most northerly East-Asian population (Yakut) having received genetic contribution from the ancestors of north European. The copying model was further used to estimate parameters of population split, which also illustrated that linkage disequilibrium pattern carried historic information beyond recent migration [35]. However, this model does not consider the later population admixture after initial population separation, which may lead to the trend of classifying the later admixed population into the population of ancestors [36].

Haplotype-sharing by accounting for linkage disequilibrium has been used to infer the human history [36, 37]. For example, haplotype analysis has been used in the study on human genetic diversity in Asia [37], in which haplotype diversity was found to be strongly correlated with latitude ($R^2 = 0.91$, $P < 0.0001$), with diversity decreasing from south to north, which consists of a loss of diversity as populations moved to higher latitudes. Phylogenic analyses of private haplotypes indicate greater similarity between East Asian and Southeast Asian, suggesting that Southeast Asia was a major geographic source of East Asian population. Another example, although it is suggested that Uyghur is a genetic donor of the East Asian [36], haplotype-sharing analysis of Uyghur showed that more than 95 % of Uyghur haplotypes could be found in either European or East Asian population, which contradicts the expectation of null hypothesis. Simulation studies further indicated that the proportion of Uyghur private haplotypes observed in the empirical data is only expected in alternative models assuming Uyghur is an admixed population.

1.3 Human Adaptation to Local Environment

There are many completely distinguished environments on the earth, varying in temperature, moisture, light, and so on. Our humans, however, have successively colonized almost every corner of the earth within only about 100,000 years since the first branch of modern humans left Africa [38]. Although the current patterns of human genetic and phenotypic diversities can, to some extent, be explained by human migration and demographic history [15, 23], natural selection and adaptive evolution have also played very important roles in different populations adapting to local environments, as revealed by modern genetics [39–41]. Investigating signatures of natural selection could facilitate identification of genes that played an important role in population adaptation to pathogens, climate, diet, and possible cognitive challenges. These studies could greatly enrich our understanding of human origins and history, and hold large potential for identifying genes with important biological functions, thus in turn, will elucidate the genetic basis of some human diseases [40, 42–45].

Although our human was not discussed in *On the Origin of Species* published in 1859 considering the religious Europe, the theory of evolution has gradually been applied to understanding human variation. As we know, a selective sweep acts on beneficial allele and makes its frequency increase quickly, and ultimately affects many loci around it due to the linkage disequilibrium. Therefore, many features of the genomic region around the target of selection are different from those evolve neutrally, which lead to the development of various statistics for detecting the natural selection. A decade ago, identifying targets of selection in humans were solely based on extremely time-consuming candidate gene studies, in which it was a priori hypothesized that a gene has been subject to natural selection and the value of statistic on it was compared with that under neutrality [46]. In addition to the inefficiency of candidate gene studies, there are many other limitations such as the need for a priori hypothesis on gene functions, difficulty in distinguishing the footprints of demographic events and inefficiency in detecting selection in regulatory elements.

Recently, the availability of genomic data offered a new paradigm for detecting signature of natural selection, which was called genomic approach. The genomic approach is not only high throughput, but also shows several statistical advantages [42, 46]. First, the genomic approach searches the genome without a priori hypothesis which gene is subjected to natural selection, essentially yielding unbiased results. Second, demographic history affects genomic loci equally while each selection event only affect a few loci, which provides the framework for distinguishing selection signals from demographic history. Third, genomic approach does not need a priori knowledge about the gene functions and genotype–phenotype relationships. The most commonly used genomic approach is the "outlier of summary statistic", calculating a statistic across all loci that quantifies some features of genetic variations, constructing empirical distribution, and defining selection candidate based on the extreme tail of the empirical distribution. The statistics of the genetic variations being calculated at genome-wide usually are allele frequency spectrum, increased derived allele frequencies (DAF), polymorphism deviating from interspecies divergence, extended haplotype homozygosity (EHH), increased population differentiation, and biased ancestry contribution in admixed populations. Tens of genome-wide studies have been conducted to detect natural selection in humans, and the initial maps of positive selection in human have been produced [42, 47]. Although these maps are incomplete, error-prone, and of low resolution, it is no doubt that this progress has fundamentally changed the field of human population genetic and evolutionary studies. Mostly importantly, the recent studies have given new insight into the genetic base of human adaptation to environment such as high latitude, high-altitude, shifted diet, and pathogens.

1.3.1 Adaptation to High-Latitude Climates

The adaptation to high latitude climates was most pronounced in the human history during/after the out of Africa [43, 48]. Skin pigmentation is the phenotype that varies most conspicuously, with darker-skinned populations concentrated in the tropics, and lighter-skinned populations in higher latitudes. In low latitude (tropic regions), the dark melanin protects people from sunburn by scattering and absorbing ultraviolet radiation; and limits photo-degradation of nutrients such as folate [49]. Therefore, any mutation that impacts normal dark melanin production is deleterious, and genes involved indark melanin production such as *MC1R* are subjected to strong purifying selection [50]. In higher latitude, with less sunlight available, depigmentation may be favored because ultraviolet penetration is necessary for vitamin D_3 synthesis. Therefore, those genes that are associated with light skin such as *SLC45A2* and *OCA2* have been subjected to recent positive selection [51]. Although polymorphisms in *ASIP* and *OCA2* played a shared role in shaping light skin [51], the evolutions of light skin color in East Asia and western Eurasia seems independent of each other and many genes underlying the environment adaptation are different [39, 40, 51]. Based on extended haplotype homozygosity (EHH) or extremely high population differentiation using high-density SNPs data, it is revealed that *SLC45A2* and *SLC24A5* have been subjected to strong positive selection and played an important role in the adaptation of the high-latitude climate in western Eurasia, while *EADR* and *ED2R* have been subjected to strong positive selection and played an important role in the adaptation of the high latitude climate in East Asia and America [40].

1.3.2 Adaptation to High Altitude

The environmental challenges for human survival and reproduction in high altitude (above 2500 m) include decreased ambient oxygen tension, increased ultraviolet radiation, extreme diurnal ranges in temperature, arid climate, and so on. However, there are approximately 140 million individuals living permanently at high altitude in North, Central, and South America, East Africa, and Asia [52, 53]. Populations living at high altitude, especially those living in the Tibetan Plateau and Andes, have evolved unique physiological characteristics compared with each other and with low-altitude populations [54, 55]. Several studies, using different technologies and strategies, have identified the genes that adapted Tibetans to the local environment based on genome-wide data [53, 56–59]. The strongest candidate genes are *EGLN1* (also known as *HIFPH2*, located in 1q42.2) and *EPAS1* (also known as *HIF2A*, located in 2p21), both of which are involved in the HIF pathway and response to hypoxia [56–58]. Xu et al. [56] analyzed the local Linkage disequilibrium of the two HIF genes and ultimately found the Tibetans' dominant haplotype. Based on the significant over-representation of the carriers of dominant haplotype of *EPAS1* and *EGLN1* in Tibetans, they proposed a "dominant haplotype carrier" model to explain the role of the two genes in adapting to high altitude [56].

1.3.3 Adaptation to Shifted Diet

Diet is one of the most important factors in species evolution and has been highlighted in *On the Origin of Species*. Facilitated by the master of fire and the recent domestication of plants and animals [60, 61], human evolution is characterized by significant dietary shifts. The human lactase persistence is one of the best examples of adaptation to shifted diet. In most humans, the capability to digest lactose, the main carbohydrate in milk, declines rapidly after weaning because the expression level of *LCT* that encode lactase-phlorizin hydrolase decreases. However, many individuals from populations with a history of milk drinking can reserve the capability to digest milk into adulthood, and the distribution of lactase persistence correlates well with that of populations with a history of cattle domestication and milk drinking [62]. The lactase persistence is inherited as a dominant Mendelian trait, and was thought to have been caused by the change of *cis* element of *LCT* [63]. A linkage disequilibrium and haplotype analysis of Finnish pedigrees has identified a causative regulatory variant (C/T-13910) ~14 kb upstream of *LCT* gene [64], with an estimated age of about 2,000–20,000 years [65]. A region of extensive linkage disequilibrium spanning >1 M has been observed in the haplotype with T-13910 allele, which is consistent with recent positive selection [65–68]. However, lactase persistent populations elsewhere such as African do not carry this variant [69, 70]. An association study on Tanzanians, Kenyans and Sudanese identified three variants (G/C-14010, T/G-13915 and C/G-13907) that could lead to lactase persistence. These mutations originated on different haplotype backgrounds from the European C/T-1390 and from each other, which indicated the independent origin of lactase persistence [70]. Further study showed that the haplotype carrying C-14010 extended >2 Mb, which is consistent with a selective sweep about 7,000 years ago. The different origins of lactase persistence also provide a perfect example of convergent evolution due to strong selective pressure as the shared dietary.

Starch consumption is a prominent characteristic of agricultural societies, especially among the populations living on planting; while rainforest and Circum-Arctic hunter-gatherers and some pastoralists consume much less starch. Interestingly,

AMY1 (human salivary amylase gene) has been reported subjected to recent positive selection in populations with planting tradition, contrasting to *LCT* in population with stockbreeding tradition [60, 69, 70], which might reflect the influence of different kinds of agriculture and culture. The copy number of the salivary amylase gene (*AMY1*) is corrected positively with the salivary amylase protein level, and individuals from population with high-starch diets [60]. Thus an individual with more copies of *AMY1* is presumably able to get more out of their starchy diet, thus providing survival advantage when food is limited. It is also suggested that higher *AMY1* copy number and protein levels might also buffer against the fitness-reducing effects of intestinal disease [60].

1.3.4 Adaptation to Pathogens and Its Impact on Human Genome

The human life expectancy was less than 25 years before Pasteur and Robert Koch discovered that microbes caused the most serious diseases. Following Pasteur's microbial theory in the later nineteenth century, the life expectancy significantly improved due to the control of infections through improved hygiene, vaccines, and antibiotics [71]. However, the relationship between pathogens and natural selection has not been realized until John B.S. Haldane investigated thalassaemia patients infecting malaria [72], which was one of the best cases showing how pathogens and natural selection are interacting and how they are shaping human genome. The genome-wide association studies are pretty powerful tools in detecting loci associated with the susceptibility or severity of infectious diseases [73, 74]. These susceptibility genes identified in this way are targets, transports, or some other components in the pathogen infectious pathway. Many studies have demonstrated the correlation between genetics variability and pathogen richness in the corresponding geographic regions [75–77]. When focusing on selective events that occurred more recently (detected by integrated haplotype score and linkage disequilibrium decay test), immunity related genes were found over-represented [68, 78]. These observations may indicate that our immune system has particularly been challenged during the recent phases of human evolution, which might propose the burden of infectious diseases that are associated with the advent of agriculture [79].

Human adaptation to pathogens is very complex and is dependent on many factors, including the temporal and spatial presence of pathogens during evolution, varying pathogenicity, host–pathogen interaction, and the rate at which pathogens evolve [76, 77, 80]. A study on the Toll-like receptor (*TLR*) gene family has concluded that viruses have exerted stronger selective pressures than other pathogens by constraining amino acid diversity at viral recognition *TLR*s [81]. Although immune-related genes played a role in protecting the host from infection, mutations inactivating those genes are likely to represent a selective advantage for the host when a pathogen uses the host immune receptors as a mechanism of cell entry and survival. Some of this kind of genes have lost their function because of the strong selective pressure, which also provided insights into the functional redundancy in our immune system [82]. Loss-of-function mutations in *CCR5*, *DARC*, *CASP12*, *SERPINA2,* and *SIGLEC12* are such cases. However, the selection may be very complex, sometimes considering the changing pathogens. For example, *CCR5*-Δ32 allele is a deletion mutation of *CCR5* that impaired the function of its coding protein and has a specific impact on the function of T cells. *CCR5*-Δ32 have been subjected to positive selection in Europe and can block the entrance of HIV-1 [83]. However, since HIV emerged very recently in Europe, the selection signals on *CCR5*-Δ32 might be caused by Black Death or/and smallpox [84]. Malaria has been, and still is, one of the major causes of child mortality in tropical regions [85]. Because of the strong selective pressure, malaria has driven most of the common monogenic diseases including sickle-cell anemia, α-thalassemia, β-thalassemia, glucose-6-phosphatase (*G6PD*) deficiency, and so on. However, these erythrocyte variants are probably only the tip of the iceberg considering all genes associated with susceptibility and resistance to malaria, many of which are involved in immune system and inflammatory genes [86]. The observations that different malaria-resistance alleles arose in different regions suggested independent evolutionary history of these genes in different parts of the world.

1.4 Human Population Admixture and Admixture Mapping

Conceptually, population admixture refers to inter-breeding among people from mutually isolated populations and giving rise to admixed population. Schematic representation was used to illustrate the salient features of population admixture (Fig. 1.1). First, two long time isolated populations contributing to the genetic information of the admixed population are referred to as parental populations. The genetic differences between the two parental populations should be significant and some markers in the genome of an individual should trace its ancestry to only one parental population. Second, in the first generation of admixtures, a hybrid infant receives half DNA from each of the two parents which is from two distinct parental populations, while each parental population contributes a haplotype to the individual. Therefore, the genome from a first-generation admixed individual comprises two haplotypes from two distinct parental populations and the genetic contribution of any parental population across each locus in the genome is exactly 50 % without any fluctuation. Third, recombination breaks and rejoins DNA molecules to form new ones in the following generations; a chromosome from an admixed population resembles a mosaic of ancestral chromosomal segments from different parental populations (Fig. 1.1). Because of the genetic drift, the ancestral proportion of a particular parental population in the admixture population across the genome shows fluctuation except the first generation.

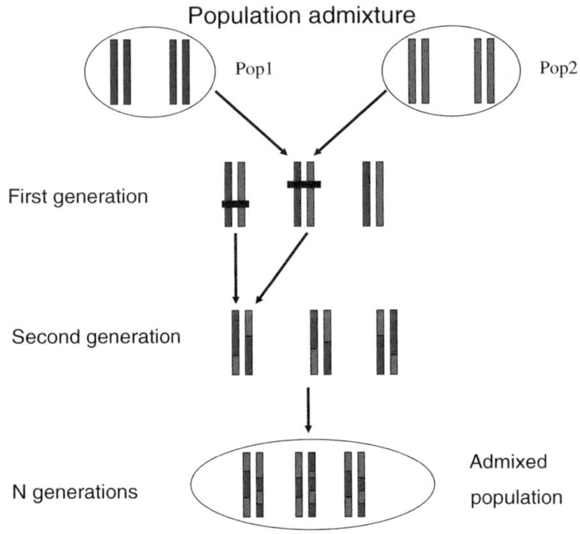

Fig. 1.1 Schematic representation of population admixture. Two parental populations, pop1 and pop2, are represented by *blue* and *red* chromosomes. The first-generation admixed individual receives a copy of haplotype from each of the parental populations with intact chromosomes. As generation is going on, the genomes of admixed population are a mosaic of the two parental chromosomes

Strictly speaking, almost all human populations showed some admixture features due to the very complex human migration and re-colonization history. However, it is very difficult to distinguish the ancestral chromosomal segments from different parental populations if the genetic distances between the two parental populations are very close or a parental population contributed very small fraction of the total DNA. Therefore, people only focus on the admixed populations with the parental populations from different continents and the admixed populations with each parental population contributing a significant fraction of the total DNA. European, African, and Asian and Amerindian met each other in the New World following the Age of Discovery, which leads to the formation of classic admixed populations such as African Americans, Mestizos, Puerto Ricans, and other Latino/Hispanic populations. These admixed populations altogether contribute a significant faction of the total population in the New World. In addition, admixed populations also distributed in other parts of the world, such as Uyghurs that are of European–Asian origin in Central Asia [36, 87, 88], and populations that are of African–Indian origin in South Asia [89, 90]. Although the population admixture itself does not create new mutations/variations, the admixed population contained much more genetic variations than any of its parental population. Especially, the novel trait combinations were created through segregation and recombination, which may lead to novel phenomena and contribute to population fitness.

Although the genetic differences between populations only represent a small fraction of total genetic variation, some diseases have different prevalence in different populations due to local adaptation or genetic drift [91, 92]. Recent researches on natural selection further showed that many genetic variants subjected to positive selection in the past thousands of years were specific to populations in particular continents [93]. Therefore, recently admixed populations

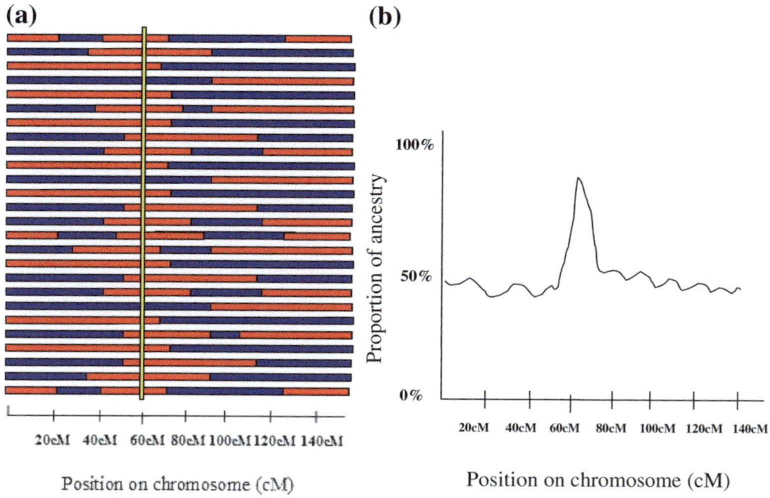

Fig. 1.2 Schematic representation of a disease locus with extreme ancestry deviation in patients. **a** Schematic of the chromosomes in patients, with *blue* and *red* representing the ancestral chromosomal segments from two distinct parental populations. **b** The ancestry proportion across the chromosome with the disease locus showing extremely deviation

such as African Americans are likely to harbor genetic variants that have important functional effects. The alleles with high-frequency difference between parental populations may become the risk allele for a disease with varying prevalence among the populations, which could be used for disease gene discovery, namely admixture mapping (Fig. 1.2). Admixture mapping has attracted many scientists in the initial stage due to its much lower cost compared with genome-wide association studies (GWAS).

Admixture mapping, in which patients are scanned for loci with extreme local ancestry deviation [94], has been investigated theoretically and has become a standard method for localizing disease causing genetic variants in admixed populations (Fig. 1.2). The statistical power of admixture mapping comes from the fact that population admixture creates linkage disequilibrium between loci with different allele frequencies in parental populations [94–96]. Since population admixture creates extended linkage disequilibrium and ancestral chromosomal segments even extend several centimorgans (cMs) in a recent admixed population, only thousands of (about 1,500–5,000) high ancestry informative markers (AIMs) will be enough for a genome-wide admixture mapping [97, 98]. Therefore, admixture mapping is very important in holding the statistical power and reducing costs [99]. Factors influencing the statistical power of admixture mapping, such as admixture dynamics, demographic history, population substructures of admixed population, have been investigated in various studies [91, 96, 100–102]. The design of admixture mapping can be either case-control or case-only, and in the latter the local ancestry of disease cases is compared with the local ancestry elsewhere in the genome.

It is of significant importance for accurate local ancestry inference because the statistical power for admixture mapping depends on it. Since AIMs are high population differentiated and unlinked in each parental population, Hidden Markov model (HMM) based approaches are implemented in several software, including ADMIXMAP, ANCESTRYAMP, and MALDsoft, to infer the local ancestry [103]. Based on these approaches, some complex diseases associated loci in African Americans have been identified [104–107]. For example, admixture mapping identified 8q24 as a prostate cancer risk locus in African–American men, which can be replicated by various GWAS [108]. Although using AIM panels is much economical, they are less powerful than that of GWAS [103], and traditional methods for local ancestry inference did not account for high-linkage disequilibrium between markers. In recent years, various methods have been developed to infer locus-specific ancestry based on high-density SNPs data, SABER [109], LAMP and LAMP-ANC [110], HAPAA [111], and HAPMIX [31]. Especially, HAPMIX employs an explicit population genetic model to infer local ancestry based on fine-scale variation data for populations formed by a two-way admixture. HAPMIX permits small rates of miscopying from the ancestral haplotype, modeling unphased diploid data from the admixed population with the HMM.

1.5 Disease-Associated Genes and Genetic Variants

Genetics plays an important role in all kinds of human diseases, either determining the occurrence or influencing the susceptibility and the severity [73, 74]. Strictly speaking, genetic diseases referred to human diseases that are caused by abnormality of individual genome. Some genetic diseases directly transmit from the parents, while other genetic diseases are caused by mutations or epigenetic change in a gene or a group of genes. Traditionally, genetic diseases are typically classified into monogenic diseases (single-gene diseases) and complex diseases [112, 113]. Monogenic diseases, such as sickle-cell anemia and cystic fibrosis, are caused by single-gene defects and their transmissions simply follow Mendel's law. Although >4,000 human monogenic diseases have been reported, these diseases altogether only affected a tiny fraction of the general population. In contrast, complex diseases, such as diabetes and cancers, are associated with the effects of multiple genes and environmental factors, and their transmissions do not show Mendelian inheritance patterns. Complex disease, also called common disease, has many health implications due to its high incidence in the general population. Elucidation of the genetic basis of the human disease is crucial for developing the suitable medicine and efficient cue approach. At the end of the twentieth century, many genes underling monogenic disorders have been identified because of the wide usage of linkage analysis on pedigree data. Although linkage studies have achieved many successes on monogenic disorders, their much-limited successes on complex diseases are attributed to their low power and low resolution for variants of modest effect. Many scientists have been continuously working on complex diseases even frustrated without powerful tools due to the health implication. With the recently development of high-throughput genotyping technology, the focus shifted swiftly to genome-wide association studies (GWAS) due to its high-throughput and high statistical power.

The allelic spectrum underlying complex disease has important influence on both research and clinical practice. In brief, it determines the strategies and methodologies for disease gene mapping. Initially, population genetics assumes a simple situation in which disease causal allele is at high frequency or intermediate frequency, but of relatively low penetrance [114, 115]. This hypothesis referred to common disease common variants (CDCV) and was the genetic basis of association studies. The international haplotype map project (HapMap) provided nearly 4 million high-frequency SNPs, genome-wide linkage disequilibrium pattern, and haplotype map [67, 116], which greatly facilitated the development of commercial high-throughput genotyping technology for GWAS. Current GWAS, in which millions of SNPs are genotyped on thousands or even tens of thousands of individuals, is a very powerful tool for investigating the genetic basis of complex diseases. Hundreds of GWAS have been conducted in identifying common variations that are statistically associated with particular diseases [116, 117]. The first wave of large-scale GWAS has improved our knowledge of genetic basis of many complex traits/diseases [118]. For example, we have witnessed rapid expansion in numbers of susceptible loci for some diseases/traits, such as type 1 diabetes, type 2 diabetes, prostate cancer, inflammatory bowel disease, breast cancer, height, fat mass, and lipid [117, 119]. Few loci have been identified for other diseases, such as asthma, coronary heart disease, and atrial fibrillation.

Most GWAS featured case-control designs, which have raised issues about the selection of suitable cases and controls. Case selection has mainly focused on the improvement of statistic power by enriching specific disease-predisposing alleles including minimizing phenotypic heterogeneity. Optimal selection of control samples remains more controversial, one economic method is to use common-control to study a series of diseases such as what Wellcome Trust Case Control Consortium (WTCCC) has done [118]. For the sample size, the consensus view is clear: the more the better. The undetected population structure (population stratification) among samples can mimic the signals of association, which increase the false positive [120–122]. Further analysis showed that: as long as cases and controls are well matched, and individual genetic outliers are excluded, the effects of residual substructure on type1 error seem modest [118]. Recently, GWAS is being extended from case-control designs to population-based cohorts which offer longitudinal measures of a wide range of quantitative traits and integrate the environmental factors for systematic analysis [119].

Overall 5,053 SNPs have been reported to be associated with hundreds of complex traits/disease by September 24, 2011 (www.genome.gov), based on a threshold of p-values $< 1.0 \times 10^{-5}$ and studies with >100,000 SNPs in the initial stage. The distribution of traits associated variants identified by GWAS is not homogeneous in the human genome [117]. It is obvious that the human HLA regions contained more complex diseases/traits associated loci. The deluge of GWAS also provided the opportunity to evaluate the potential impact of genetic variants on complex diseases by systematically cataloging and summarizing the characteristics of the identified trait/disease-associated SNPs (TASs). Unsurprisingly, since GWAS was primarily powered for common variants, risk allele frequencies were well above 5 % (median risk allele frequency 36 %, interquartile range 21–53 %) in the populations analyzed [117]. Analyses of 465 unique TASs showed that 43 % TASs were located in intergenic regions, 45 % were intronic, 9 % were non-synonymous, 2 % were located in a 5′ UTR or 3′ UTR, and 2 % were synonymous [117]. The odds ratios of discrete traits ranged from 1.04 to 29.4 (median 1.33, interquartile range 1.20–1.61). Evolutionary analysis of TASs showed these SNPs were subjected to a slightly high positive selection compared with that of background.

Although GWAS is a great success, all the variants having been identified only accounted for a small increment in risk and explained a small fraction of estimated heritability. For example, human height is a classic complex trait with an estimated heritability of about 80 %, however, more than 40 associated loci identified by GWAS explain only about 5 % of phenotypic variance based tens of thousands of individuals [123]. Many assumptions have been proposed to explain the missing heritability, including much more variants of smaller effect that have not been found, rare variants that are not examined in the commercial chips, undetected epistasis, epigenetic effect, and inadequate accounting for shared environment among relatives [124]. The missing heritability in GWAS led people to prefer the alternative hypothesis—common disease rare variants (CDRV), which argues that the major genetic susceptibility to the complex disease are multiple rare variations with high penetrance [125, 126].

The next generation sequencing has the potential to discover the entire allelic spectrum of variants in well-phenotyped individuals. Especially, deep sequencing approaches have been proved to be successful in the study of monogenic disorders [127, 128]. However, the applications of the next generation sequencing on complex diseases studies, which generally require sequencing thousands of individuals, remain to be a challenge due to the high costs and limits of existing sequencing capacity. Three strategies have been proposed to take advantage of the next generation sequencing: imputation, genotyping, and low-coverage sequencing [128, 129]. First, imputing previously genotyped samples using the recent sequenced reference panel is the most economic way, albeit less accurate [129]. Second, commercial chips integrating the new discovered SNPs would essentially improve the statistic power in identifying the disease-associated variants. Third, low-coverage sequencing (2–6 × coverage) of many individuals is the economic way for association study [128]. Since next generation sequencing detects millions of rare variants, these data have three features: high proportion of rare variants, high error, and high missing data [130]. However, traditional statistics testing the association of common alleles with complex diseases are not suitable for large amount of allelic heterogeneity presenting in sequence data [131] and do not account for the large amounts of sequence errors or missing data. Various association tests have been proposed to analyze these coming data with new features [130, 132–134]. For examples, Li and Leal [132] developed a combined multivariate and collapsing (CMC) method taking advantage of both collapsing and multiple-marker tests, and demonstrated that CMC was both powerful and robust using sequencing data.

References

1. McHenry HM. Human evolution. In: Travis MRJ, editor. Evolution: the first four billion years. Cambridge: The Belknap Press of Harvard University Press; 2009. p. 265.
2. Dart R. Australopithecus africanus: the man-ape of South Africa. Nature. 1925;115:195–9.
3. Stauffer RL, Walker A, Ryder OA, Lyons-Weiler M, Hedges SB. Human and ape molecular clocks and constraints on paleontological hypotheses. J Hered. 2001;92:469–74.
4. Wood B. Origin and evolution of the genus Homo. Nature. 1992;355:783–90.
5. Ke Y, et al. African origin of modern humans in East Asia: a tale of 12,000 Y chromosomes. Science. 2001;292:1151–3.
6. Ingman M, Kaessmann H, Paabo S, Gyllensten U. Mitochondrial genome variation and the origin of modern humans. Nature. 2000;408:708–13.
7. Cann RL, Stoneking M, Wilson AC. Mitochondrial DNA and human evolution. Nature. 1987;325:31–6.
8. Vigilant L, Stoneking M, Harpending H, Hawkes K, Wilson AC. African populations and the evolution of human mitochondrial DNA. Science. 1991;253:1503–7.
9. Green RE, et al. A draft sequence of the Neandertal genome. Science. 2010;328:710–22.
10. Reich D, et al. Genetic history of an archaic hominin group from Denisova Cave in Siberia. Nature. 2010;468:1053–60.
11. Reich D, et al. Denisova admixture and the first modern human dispersals into Southeast Asia and Oceania. Am J Hum Genet. 2011;89:516–28.
12. Rasmussen M, et al. An aboriginal Australian genome reveals separate human dispersals into Asia. Science. 2011;333:94–8.
13. Hey J, Machado CA. The study of structured populations–new hope for a difficult and divided science. Nat Rev Genet. 2003;4:535–43.
14. Ramachandran S, et al. Support from the relationship of genetic and geographic distance in human populations for a serial founder effect originating in Africa. Proc Natl Acad Sci USA. 2005;102:15942–7.
15. Li JZ, et al. Worldwide human relationships inferred from genome-wide patterns of variation. Science. 2008;319:1100–4.
16. Cann HM, et al. A human genome diversity cell line panel. Science. 2002;296:261–2.
17. Cavalli-Sforza LL. The human genome diversity project: past, present and future. Nat Rev Genet. 2005;6:333–40.
18. Falush D, Stephens M, Pritchard JK. Inference of population structure using multilocus genotype data: linked loci and correlated allele frequencies. Genetics. 2003;164:1567–87.
19. Pritchard JK, Stephens M, Donnelly P. Inference of population structure using multilocus genotype data. Genetics. 2000;155:945–59.
20. Rosenberg NA, et al. Genetic structure of human populations. Science. 2002;298:2381–5.
21. Tang H, Peng J, Wang P, Risch NJ. Estimation of individual admixture: analytical and study design considerations. Genet Epidemiol. 2005;28:289–301.
22. Pool JE, Hellmann I, Jensen JD, Nielsen R. Population genetic inference from genomic sequence variation. Genome Res. 2010;20:291–300.
23. Rosenberg NA, et al. Genotype, haplotype and copy-number variation in worldwide human populations. Nature. 2008;451:998–1003.

24. Menozzi P, Piazza A, Cavalli-Sforza L. Synthetic maps of human gene frequencies in Europeans. Science. 1978;201:786–92.
25. Patterson N, Price AL, Reich D. Population structure and eigenanalysis. PLoS Genet. 2006;2:e190.
26. Reich D, Price AL, Patterson N. Principal component analysis of genetic data. Nat Genet. 2008;40:491–2.
27. Semino O, et al. Origin, diffusion, and differentiation of Y-chromosome haplogroups E and J: inferences on the neolithization of Europe and later migratory events in the Mediterranean area. Am J Hum Genet. 2004;74:1023–34.
28. Novembre J, Stephens M. Interpreting principal component analyses of spatial population genetic variation. Nat Genet. 2008;40:646–9.
29. Novembre J, et al. Genes mirror geography within Europe. Nature. 2008;456:274.
30. McVean G. A genealogical interpretation of principal components analysis. Plos Genet. 2009;5.
31. Price AL, et al. Sensitive detection of chromosomal segments of distinct ancestry in admixed populations. PLoS Genet. 2009;5:e1000519.
32. Pool JE, Nielsen R. Inference of historical changes in migration rate from the lengths of migrant tracts. Genetics. 2009;181:711–9.
33. Li N, Stephens M. Modeling linkage disequilibrium and identifying recombination hotspots using single-nucleotide polymorphism data. Genetics. 2003;165:2213–33.
34. Hellenthal G, Auton A, Falush D. Inferring human colonization history using a copying model. PLoS Genet. 2008;4:e1000078.
35. Davison D, Pritchard JK, Coop G. An approximate likelihood for genetic data under a model with recombination and population splitting. Theor Popul Biol. 2009;75:331–45.
36. Xu S, Jin W, Jin L. Haplotype-sharing analysis showing Uyghurs are unlikely genetic donors. Mol Biol Evol. 2009;26:2197–206.
37. HUGO Pan-Asian SNP Consortium et al. Mapping human genetic diversity in Asia. Science. 2009;326:1541–5.
38. Tattersall I. Human origins: Out of Africa. Proc Natl Acad Sci USA. 2009;106:16018–21.
39. Pickrell JK, et al. Signals of recent positive selection in a worldwide sample of human populations. Genome Res. 2009;19:826–37.
40. Sabeti PC, et al. Genome-wide detection and characterization of positive selection in human populations. Nature. 2007;449:913–8.
41. Coop G, et al. The role of geography in human adaptation. PLoS Genet. 2009;5:e1000500.
42. Akey JM. Constructing genomic maps of positive selection in humans: where do we go from here? Genome Res. 2009;19:711–22.
43. Balaresque PL, Ballereau SJ, Jobling MA. Challenges in human genetic diversity: demographic history and adaptation. Hum Mol Genet. 2007;16(R2):R134–9.
44. Nielsen R, Hellmann I, Hubisz M, Bustamante C, Clark AG. Recent and ongoing selection in the human genome. Nat Rev Genet. 2007;8:857–68.
45. Hancock AM, et al. Adaptations to climate in candidate genes for common metabolic disorders. PLoS Genet. 2008;4:e32.
46. Sabeti PC, et al. Positive natural selection in the human lineage. Science. 2006;312:1614–20.
47. Novembre J, Di Rienzo A. Spatial patterns of variation due to natural selection in humans. Nat Rev Genet. 2009;10:745–55.
48. Pritchard JK, Pickrell JK, Coop G. The genetics of human adaptation: hard sweeps, soft sweeps, and polygenic adaptation. Curr Biol. 2010;20:R208–15.
49. Jablonski NG, Chaplin G. Human skin pigmentation as an adaptation to UV radiation. Proc Natl Acad Sci USA. 2010;107:8962–8.
50. Harding RM, et al. Evidence for variable selective pressures at MC1R. Am J Hum Genet. 2000;66:1351–61.
51. Norton HL, et al. Genetic evidence for the convergent evolution of light skin in Europeans and East Asians. Mol Biol Evol. 2007;24:710–22.

52. Moore LG. Human genetic adaptation to high altitude. High Alt Med Biol. 2001;2:257–79.
53. Bigham A et al. identifying signatures of natural selection in Tibetan and andean populations using dense genome scan data. Plos Genet. 2010;6.
54. Baker PT. Human adaptation to high altitude. Science. 1969;163:1149–000.
55. Beall CM, et al. Hemoglobin concentration of high-altitude Tibetans and Bolivian Aymara. Am J Phys Anthropol. 1998;106:385–400.
56. Xu S, et al. A genome-wide search for signals of high-altitude adaptation in Tibetans. Mol Biol Evol. 2011;28:1003–11.
57. Yi X, et al. Sequencing of 50 human exomes reveals adaptation to high altitude. Science. 2010;329:75–8.
58. Simonson TS, et al. Genetic evidence for high-altitude adaptation in Tibet. Science. 2010;329:72–5.
59. Aggarwal S, et al. EGLN1 involvement in high-altitude adaptation revealed through genetic analysis of extreme constitution types defined in Ayurveda. Proc Natl Acad Sci USA. 2010;107:18961–6.
60. Perry GH, et al. Diet and the evolution of human amylase gene copy number variation. Nat Genet. 2007;39:1256–60.
61. Diamond J. Evolution, consequences and future of plant and animal domestication. Nature. 2002;418:700–7.
62. Swallow DM. Genetics of lactase persistence and lactose intolerance. Annu Rev Genet. 2003;37:197–219.
63. Wang YX, et al. The lactase persistence/non-persistence polymorphism is controlled by a cis-acting element. Hum Mol Genet. 1995;4:657–62.
64. Peltonen L, et al. Identification of a variant associated with adult-type hypolactasia. Nat Genet. 2002;30:233–7.
65. Bersaglieri T, et al. Genetic signatures of strong recent positive selection at the lactase gene. Am J Hum Genet. 2004;74:1111–20.
66. Poulter M, et al. The causal element for the lactase persistence/non-persistence polymorphism is located in a 1 Mb region of linkage disequilibrium in Europeans. Ann Hum Genet. 2003;67:298–311.
67. Altshuler D, et al. A haplotype map of the human genome. Nature. 2005;437:1299–320.
68. Voight BF, Kudaravalli S, Wen XQ, Pritchard JK. A map of recent positive selection in the human genome. PLoS Biol. 2006;4:446–58.
69. Swallow DM, et al. A novel polymorphism associated with lactose tolerance in Africa: multiple causes for lactase persistence? Hum Genet. 2007;120:779–88.
70. Tishkoff SA, et al. Convergent adaptation of human lactase persistence in Africa and Europe. Nat Genet. 2007;39:31–40.
71. Casanova JL, Abel L. Inborn errors of immunity to infection: the rule rather than the exception. J Exp Med. 2005;202:197–201.
72. Haldane JBS. Disease and evolution (Reprinted from La Ricerca Scientifica Supplemento, Vol 19, pp 1–11, 1949). Curr Sci. 1992;63:599–604.
73. Davila S, et al. Genetic association and expression studies indicate a role of toll-like receptor 8 in pulmonary tuberculosis. PLoS Genet. 2008;4:e1000218.
74. Fellay J, et al. A whole-genome association study of major determinants for host control of HIV-1. Science. 2007;317:944–7.
75. Sironi M, et al. Widespread balancing selection and pathogen-driven selection at blood group antigen genes. Genome Res. 2009;19:199–212.
76. Prugnolle F, et al. Pathogen-driven selection and worldwide HLA class I diversity. Curr Biol. 2005;15:1022–7.
77. Sironi M, et al. Parasites represent a major selective force for interleukin genes and shape the genetic predisposition to autoimmune conditions. J Exp Med. 2009;206:1395–408.
78. Moyzis RK, Wang ET, Kodama G, Baidi P. Global landscape of recent inferred Darwinian selection for Homo sapiens. Proc Natl Acad Sci USA. 2006;103:135–40.

79. Wolfe ND, Dunavan CP, Diamond J. Origins of major human infectious diseases. Nature. 2007;447:279–83.
80. Barreiro LB, Quintana-Murci L. From evolutionary genetics to human immunology: how selection shapes host defence genes. Nat Rev Genet. 2010;11:17–30.
81. Barreiro LB et al. Evolutionary dynamics of human toll-like receptors and their different contributions to host defense. Plos Genet. 2009;5.
82. Casanova JL, Quintana-Murci L, Alcais A, Abel L. Immunology in natura: clinical, epidemiological and evolutionary genetics of infectious diseases. Nat Immunol. 2007;8:1165–71.
83. Arenzana-Seisdedos F, Parmentier M. Genetics of resistance to HIV infection: role of co-receptors and co-receptor ligands. Semin Immunol. 2006;18:387–403.
84. Galvani AP, Slatkin M. Evaluating plague and smallpox as historical selective pressures for the CCR5-Delta 32 HIV-resistance allele. Proc Natl Acad Sci USA. 2003;100:15276–9.
85. Snow RW, Guerra CA, Noor AM, Myint HY, Hay SI. The global distribution of clinical episodes of Plasmodium falciparum malaria. Nature. 2005;434:214–7.
86. Kwiatkowski DP. How malaria has affected the human genome and what human genetics can teach us about malaria. Am J Hum Genet. 2005;77:171–92.
87. Xu S, Huang W, Qian J, Jin L. Analysis of genomic admixture in Uyghur and its implication in mapping strategy. Am J Hum Genet. 2008;82:883–94.
88. Xu S, Jin L. A genome-wide analysis of admixture in Uyghurs and a high-density admixture map for disease-gene discovery. Am J Hum Genet. 2008;83:322–36.
89. Narang A, et al. Recent admixture in an Indian population of African ancestry. Am J Hum Genet. 2011;89:111–20.
90. Shah AM, et al. Indian Siddis: African descendants with Indian admixture. Am J Hum Genet. 2011;89:154–61.
91. Smith MW, O'Brien SJ. Mapping by admixture linkage disequilibrium: advances, limitations and guidelines. Nat Rev Genet. 2005;6:623–32.
92. Bamshad M, Wooding S, Salisbury BA, Stephens JC. Deconstructing the relationship between genetics and race. Nat Rev Genet. 2004;5:598–609.
93. Grossman SR, et al. A composite of multiple signals distinguishes causal variants in regions of positive selection. Science. 2010;327:883–6.
94. Chakraborty R, Weiss KM. Admixture as a tool for finding linked genes and detecting that difference from allelic association between loci. Proc Natl Acad Sci USA. 1988;85:9119–23.
95. Stephens JC, Briscoe D, O'Brien SJ. Mapping by admixture linkage disequilibrium in human populations: limits and guidelines. Am J Hum Genet. 1994;55:809–24.
96. Pfaff CL, et al. Population structure in admixed populations: effect of admixture dynamics on the pattern of linkage disequilibrium. Am J Hum Genet. 2001;68:198–207.
97. Smith MW, et al. A high-density admixture map for disease gene discovery in african americans. Am J Hum Genet. 2004;74:1001–13.
98. Tian C, et al. A genomewide single-nucleotide-polymorphism panel with high ancestry information for African American admixture mapping. Am J Hum Genet. 2006;79:640–9.
99. Xu S, et al. Dissecting linkage disequilibrium in african-american genomes: roles of markers and individuals. Mol Biol Evol. 2007;24:2049–58.
100. Pfaff CL, Kittles RA, Shriver MD. Adjusting for population structure in admixed populations. Genet Epidemiol. 2002;22:196–201.
101. Seldin MF, et al. Putative ancestral origins of chromosomal segments in individual african americans: implications for admixture mapping. Genome Res. 2004;14:1076–84.
102. Pritchard JK, Donnelly P. Case-control studies of association in structured or admixed populations. Theor Popul Biol. 2001;60:227–37.
103. Seldin MF, Pasaniuc B, Price AL. New approaches to disease mapping in admixed populations. Nat Rev Genet. 2011;12:523–8.
104. Zhu X, et al. Admixture mapping for hypertension loci with genome-scan markers. Nat Genet. 2005;37:177–81.

105. Freedman ML, et al. Admixture mapping identifies 8q24 as a prostate cancer risk locus in African-American men. Proc Natl Acad Sci USA. 2006;103:14068–73.
106. Cheng C-Y, et al. Admixture mapping of 15,280 African Americans identifies obesity susceptibility loci on chromosomes 5 and X. PLoS Genet. 2009;5:e1000490.
107. Reich D, et al. A whole-genome admixture scan finds a candidate locus for multiple sclerosis susceptibility. Nat Genet. 2005;37:1113–8.
108. Gudmundsson J, et al. Genome-wide association study identifies a second prostate cancer susceptibility variant at 8q24. Nat Genet. 2007;39:631–7.
109. Tang H, Coram M, Wang P, Zhu X, Risch N. Reconstructing genetic ancestry blocks in admixed individuals. Am J Hum Genet. 2006;79:1–12.
110. Sankararaman S, Sridhar S, Kimmel G, Halperin E. Estimating local ancestry in admixed populations. Am J Hum Genet. 2008;82:290–303.
111. Sundquist A, Fratkin E, Do CB, Batzoglou S. Effect of genetic divergence in identifying ancestral origin using HAPAA. Genome Res. 2008;18:676–82.
112. Antonarakis SE, Chakravarti A, Cohen JC, Hardy J. Mendelian disorders and multifactorial traits: the big divide or one for all? Nat Rev Genet. 2010;11:380–4.
113. Sidransky E. Heterozygosity for a Mendelian disorder as a risk factor for complex disease. Clin Genet. 2006;70:275–82.
114. Di Rienzo A, Hudson RR. An evolutionary framework for common diseases: the ancestral-susceptibility model. Trends Genet. 2005;21:596–601.
115. Di Rienzo A. Population genetics models of common diseases. Curr Opin Genet Dev. 2006;16:630–6.
116. Frazer KA, et al. A second generation human haplotype map of over 3.1 million SNPs. Nature. 2007;449:851–61.
117. Hindorff LA, et al. Potential etiologic and functional implications of genome-wide association loci for human diseases and traits. Proc Natl Acad Sci USA. 2009;106:9362–7.
118. Genome-wide association study of 14,000 cases of seven common diseases and 3,000 shared controls. Nature. 2007;447:661–78.
119. McCarthy MI, et al. Genome-wide association studies for complex traits: consensus, uncertainty and challenges. Nat Rev Genet. 2008;9:356–69.
120. Price AL, et al. Principal components analysis corrects for stratification in genome-wide association studies. Nat Genet. 2006;38:904–9.
121. Clayton DG, et al. Population structure, differential bias and genomic control in a large-scale, case-control association study. Nat Genet. 2005;37:1243–6.
122. Marchini J, Cardon LR, Phillips MS, Donnelly P. The effects of human population structure on large genetic association studies. Nat Genet. 2004;36:512–7.
123. Visscher PM. Sizing up human height variation. Nat Genet. 2008;40:489–90.
124. Manolio TA, et al. Finding the missing heritability of complex diseases. Nature. 2009;461:747–53.
125. Pritchard JK. Are rare variants responsible for susceptibility to complex diseases? Am J Hum Genet. 2001;69:124–37.
126. Kryukov GV, Pennacchio LA, Sunyaev SR. Most rare missense alleles are deleterious in humans: implications for complex disease and association studies. Am J Hum Genet. 2007;80:727–39.
127. Ng SB, et al. Targeted capture and massively parallel sequencing of 12 human exomes. Nature. 2009;461:272–6.
128. Li Y, Sidore C, Kang HM, Boehnke M, Abecasis GR. Low-coverage sequencing: implications for design of complex trait association studies. Genome Res. 2011;21:940–51.
129. Durbin RM, et al. A map of human genome variation from population-scale sequencing. Nature. 2010;467:1061–73.
130. Luo L, Boerwinkle E, Xiong M. Association studies for next-generation sequencing. Genome Res. 2011;21:1099–108.

131. Gorlov IP, Gorlova OY, Sunyaev SR, Spitz MR, Amos CI. Shifting paradigm of association studies: value of rare single-nucleotide polymorphisms. Am J Hum Genet. 2008;82:100–12.
132. Li B, Leal SM. Methods for detecting associations with rare variants for common diseases: application to analysis of sequence data. Am J Hum Genet. 2008;83:311–21.
133. Li Y, Byrnes AE, Li M. To identify associations with rare variants, just WHaIT: weighted haplotype and imputation-based tests. Am J Hum Genet. 2010;87:728–35.
134. Price AL, et al. Pooled association tests for rare variants in exon-resequencing studies. Am J Hum Genet. 2010;86:832–8.

Chapter 2
Distribution of Length of Ancestral Chromosomal Segments in Admixed Genomes

Abstract The ancestral chromosomal segments in admixed genome are of significant importance for both population history inference and admixture mapping because they essentially provide the basic information for tracking population genetic events. We introduced a theoretical framework on the distributions of the length of ancestral chromosomal segments (LACS) in hybrid isolation (HI) model and gradual admixture (GA) model. Although the distribution of LACS in GA model differs from that in HI model, we demonstrated that the mean LACS in HI model is approximately half of that in GA model if both admixture proportion and admixture time in the two models are identical. In addition, we found the peak of association signatures in HI model was much narrower and sharper than that in GA model, indicating that the identification of putative causal allele in HI model is more efficient than that in GA model.

Keywords Population admixture · Ancestral chromosomal segment · Hybrid isolation (HI) model · Gradual admixture (GA) model · Admixture mapping

2.1 Ancestral Chromosomal Segments in Admixed Genome

Admixed populations come into being when two or more previously mutually isolated populations are inter-mating. Human diasporas over the past millennium have resulted in more frequent inter-continental marriages thus more population admixture. Following the Age of Discovery, population admixture has been a common phenomenon as previously isolated populations come into contact through colonization and migration [1–5]. The recently admixed populations, such as African Americans and Mestizos, have received much attention because of their

W. Jin, *Admixture Dynamics, Natural Selection and Diseases in Admixed Populations*, Springer Theses, DOI 10.1007/978-94-017-7408-6_2

potential advantages in the discovery of disease-associated genes. Specifically, a disease gene mapping strategy named admixture mapping has been developed [6–9]. In fact, accurate understanding of population admixture dynamics is not only important to admixture mapping but also to other applications, such as elucidating population history [10] and detecting natural selection signatures in admixed populations [11, 12].

Since recombination breaks and rejoins DNA molecules to form new ones, a chromosome from an admixed population resembles a mosaic of ancestral chromosomal segments from distinct parental populations. Furthermore, distribution of ancestral chromosomal segments can be reshaped and rearranged by recombination per generation, which essentially provides valuable information about the population history [10, 13]. In short, the ancestral chromosomal segments from different parental populations are spliced into shorter pieces as the number of generation increases, while the ancestral chromosomal segments in the admixed population with a few generations are usually much longer. With the availability of high-density single nucleotide polymorphisms (SNPs) data in recent years, it is now feasible to infer population history based on the length of ancestral chromosomal segments (LACS). In addition, a number of methods and software have been developed to identify ancestral chromosomal segments based on the high-density genomic data [2, 14–16]. Application of these methods to the empirical data has significantly increased our knowledge about population history [13, 14, 17, 18]. However, most of these studies relied on simplified models that do not take into account the inherent complexity of the admixture process.

2.2 Two Classic Admixture Models

In reality, population admixture processes are either unknown or too complex to study directly. In this study, we attempted to infer the distribution of LACS in two classic admixture models representing two extreme cases of population admixture (Fig. 2.1): hybrid isolation (HI) model [12] and gradual admixture (GA) model [19]. A number of previous studies have used the two models to examine the genetic structure and linkage disequilibrium pattern of admixed populations [12, 19–21]. We simply assumed two parental populations, pop1 and pop2, contribute all the genetic material to the admixed population in both models, in which m and $1 - m$ represent the final genetic contribution of the parental populations, pop1 and pop2, to the admixed population. T represents the final number of generations since the first admixture, while t represents the number of generations the admixed population has experienced, ranging from 1 to T. In the HI model, admixture occurs only in the first generation and there is no further genetic contribution from the parental populations (Fig. 2.1a). In the following generation, the admixed population only experiences recombination and genetic drift. In the GA model, admixture occurs gradually and the parental populations contribute gene flow to the admixed population per generation. The genetic contributions of parental

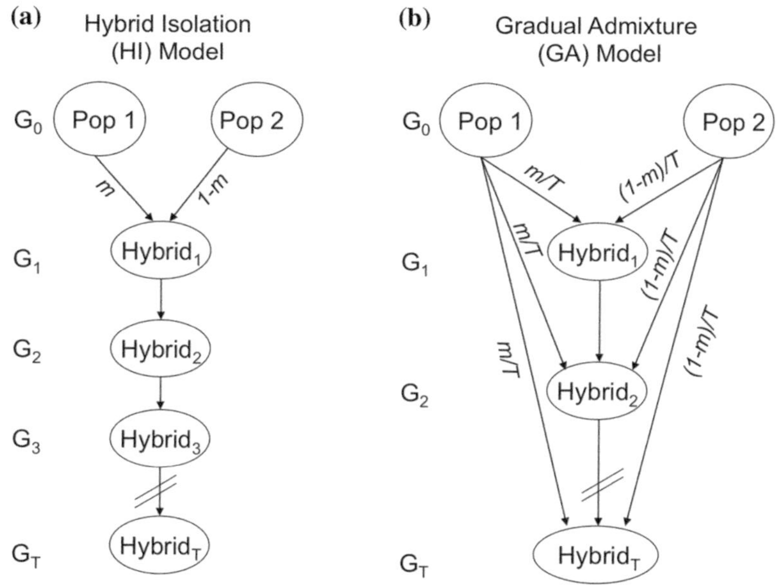

Fig. 2.1 Schematic diagram of Hybrid isolation model (**a**) and Gradual admixture model (**b**) to infer the distribution of ancestral chromosomal segments. Hybrid isolation (HI) model and gradual admixture (GA) model were adopted from previous studies [12, 19]. In each model, the genetic contributions of pop1 and pop2 were m and $1 - m$, respectively. The admixed population experienced t generations ranging from 1 to T generation. Reprint from Jin et al. [24]

populations pop1 and pop2 to the admixed population per generation are m/T and $(1 - m)/T$, respectively (Fig. 2.1b). This ensures that the genetic contribution of a given parental population to the admixed population is constant in each generation.

2.3 Distribution of LACS in HI Model

In HI model (Fig. 2.1a) the two parental populations, pop1 and pop2, inter-mated T generations ago and their genetic contributions to the admixed population were m and $1 - m$, respectively. We used genetic distance (Morgan) to measure the length of ancestral chromosomal segments (LACS). Recombination events occurring along the chromosome would follow a Poisson process at a rate of 1 per generation if the ends of the chromosomes were ignored. Recombination occurring between two ancestral chromosomal segments, both from pop1, could be ignored if the genetic contribution of pop1 (m) was very small. For a particular chromosome from pop1, each recombination event would cut the ancestral chromosomal segments into smaller pieces. As a result, the ancestral chromosomal segments entering the admixed population T generations ago followed an exponential distribution with mean $1/t$: $f(x;\ T) = te^{-Tx}$ [10].

However, m is usually not a small value in classic admixture models and influences of recombination on ancestral chromosomal segments from the same parental population should not be ignored. We treated pop1 as the given parental population and analyzed the ancestral chromosomal segments from it to illustrate the distribution of LACS. The probability that a given ancestral segment from pop1 could recombine with those from the same parental population was m. However, recombination among ancestral chromosomal segments from the same parental population could not change the LACS distribution, while recombination among chromosomal segments from distinct parental populations tends to cut the segments into smaller pieces. The probability that a particular ancestral segment from pop1 recombined with ancestral chromosomal segments from pop2 was $1 - m$. Thus, chromosome segments from pop1 recombining with those from pop2 followed a Poisson process at a rate of $1 - m$. Then, after T generations, the distribution of LACS from pop1 in the admixed population would follow an exponential distribution with mean $\frac{1}{(1-m)T}$, as shown below.

$$f(x; T) = (1 - m)Te^{-(1-m)Tx} \tag{2.1}$$

where distribution of LACS in HI model was influenced by the genetic contribution of parental population (m) and the number of generations (T). In short, the lesser the genetic contribution from parental population (m), the shorter the ancestral chromosomal segments; the larger the number of generations (T), the shorter the ancestral chromosomal segments.

2.4 Distribution of LACS in GA Model

In GA (Fig. 2.1b) model, parental populations gradually contributed their genetic material to the admixed population over T generations. Relative gene flow from parental populations, pop1 and pop2, to the admixed population at each generation was m/T and $(1 - m)/T$, respectively, with rest of the genetic materials from previous generations of the admixed population. Therefore, the final genetic contributions of pop1 and pop2 to the admixed population were m and $1 - m$, respectively. Similar to the HI model, we treated pop1 as the given parental population and analyzed the ancestral chromosomal segments from pop1 to illustrate the distribution of LACS. If a chromosome from pop1 entered the admixed population t generations ago, the distribution of LACS derived from this chromosome would be the same as that in HI model: $f(x; t) = (1 - m)te^{-(1-m)tx}$. If chromosome ends were ignored, a chromosome from pop1 would be expected to split into $E[k(t)] = (1 - m)t$ pieces per unit length. Then, the contribution of ancestral segments from pop1 to the admixed population is proportional to $(1 - m)t$ after t generations. Therefore, ancestral segments from pop1 included segments from different times over T generations,

$$\begin{aligned} f(x;T) &= \frac{\int_0^T (1-m)te^{-(1-m)tx}(1-m)t\mathrm{d}t}{\int_0^T (1-m)t\mathrm{d}t} \\ &= \frac{4-2e^{-(1-m)Tx}\left((1-m)^2T^2x^2+2T(1-m)x+2\right)}{(1-m)^2T^2x^3} \end{aligned} \tag{2.2}$$

where distribution of LACS in GA model was observed to be influenced by the genetic contribution of parental population (m) and the number of generations (T).

2.5 Consistency Between Theoretical and Simulated Distributions

We simulated data under several scenarios with forward-time simulation program. The genetic contribution of the parental population to the admixed population (m) was set at 50 %. The effective population sizes (N_e) of parental and admixed populations were set at 5,000, although it would not influence the distribution. The number of generations T was set to 10, 20, 50, and 100, respectively. The theoretical distributions of LACS with identical parameters to the simulated data were obtained for a comparative analysis. As shown, the theoretical distribution was consistent with the simulated distribution in both HI and GA models in all scenarios (Fig. 2.2a, b). Further analysis showed no significant difference between theoretical and simulated distributions ($P > 0.05$, Kolmogorov–Smirnov tests [22]). Although theoretical distribution of LACS with 10 generations essentially fit the simulated data, further analysis showed that the simulated data lacked long ancestral chromosomal segments compared to theoretical distribution. These differences were essentially caused by theoretical distribution assuming an infinite chromosome length, while the simulated data were based on real length of chromosomes with fixed and finite lengths. With $t = 1$, LACS in theoretical distribution was ill defined and the distribution becomes more accurate when t became larger. We also found that the distributions of LACS among different generations were significantly different from each other in each tested model ($P < 10^{-16}$, Kolmogorov–Smirnov tests [22]). The difference of distributions observed among different generations was more pronounced when the 10-based logarithm of LACS was obtained (Fig. 2.2c, d).

2.6 Comparison of Distribution of LACS Between HI and GA Models

Since the ancestral chromosomal segments in HI model followed an exponential distribution, the expectation of LACS from pop1 at generation T was

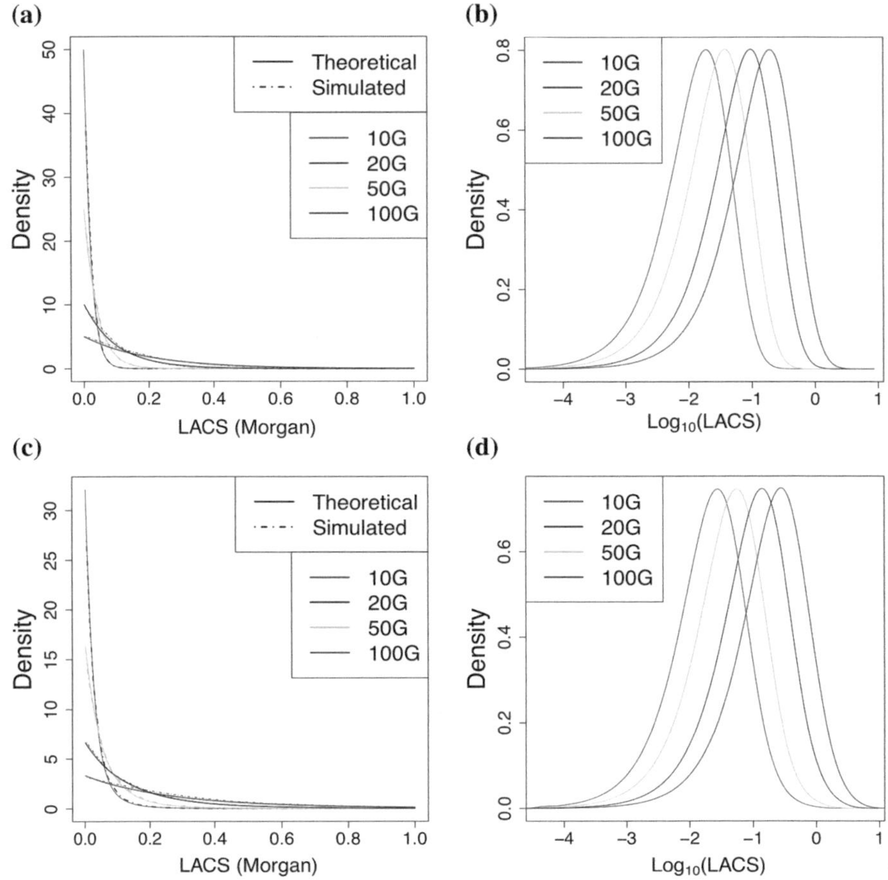

Fig. 2.2 Distribution of the length of ancestral chromosomal segments (LACS) in HI and GA models. The genetic contribution from parental populations to the admixed population was assumed to be 50 %. The number of generations since admixture was assumed to be 10, 20, 50, and 100, respectively. G denotes the number of generations since admixture. *Solid lines* represent theoretical distribution and *dashed lines* represent simulated distribution. Distribution of LACS in HI model (**a**) and GA (**c**) model. Distribution of Log_{10} (LACS) in HI model (**b**) and GA (**d**) model. Reprint from Jin et al. [24]

$$E(x; T) = \frac{1}{(1-m)T} \tag{2.3}$$

and the variance was

$$D(x; T) = \frac{1}{(1-m)^2 T^2} \tag{2.4}$$

In T-generation GA model, the expected mean and variance of LACS could be calculated based on the definition: $E(x; T) = \int_0^\infty x f(x; t) \mathrm{d}x$ and

$D(x;T) = \int_0^\infty (x - E(x;T))^2 f(x;T)\mathrm{d}x$. To our knowledge, it is impossible to integrate the formula directly to obtain an analytic expression. Therefore, we used an alternate approach to estimate the mean and variance of LACS in GA model. Based on the method to calculate density function, we could see that the mean LACS from pop1 could be calculated by averaging the mean LACS from different timescales. Since ancestral chromosomal segments from different generations have different weights and the weight $P(\overline{x_t})$ is proportional to $w_t = (1-m)t$, the mean $\overline{x_t}$ from different times corrected by weights could be calculated as:

$$\hat{E}(x;T) = \int_0^T \overline{x_t} P(\overline{x_t})\mathrm{d}t = \frac{\int_0^T \overline{x_t} w_t}{\int_0^T w_t} = \frac{\int_0^T \frac{1}{(1-m)t}(1-m)t}{\int_0^T (1-m)t} = \frac{2}{(1-m)T} \tag{2.5}$$

Similar strategies could be applied to estimate the variance of LACS in the GA model.

$$\begin{aligned}\hat{D}(x;T) &= E(x^2;T) - E^2(x;T) = \sum_{t=1}^{t=T} p_t E(x_t^2) - \left(\frac{2}{(1-m)T}\right)^2 \\ &= \frac{\sum_{t=1}^{t=T}(1-m)tE(x_t^2)}{\sum_{t=1}^{t=T}(1-m)t} - \left(\frac{2}{(1-m)T}\right)^2 \\ &= \frac{\sum_{t=1}^{t=T}(1-m)t(D(x_t) + E^2(x_t))}{\sum_{t=1}^{t=T}(1-m)t} - \left(\frac{2}{(1-m)T}\right)^2 \\ &\approx \frac{4(\sum_{t=1}^{t=T}\frac{1}{t} - 1)}{(1-m)^2T^2}\end{aligned} \tag{2.6}$$

Although we deduced the theoretical distribution of ancestral chromosomal segments in both HI and GA models separately, we were also interested in the relationship between these distributions. Based on (2.3) and (2.5), we could observe that the mean LACS of a given parental population in GA model is twice as that in HI model if both the number of generations (T) and genetic contribution (m) in the two models were identical, which also indicated that the mean LACS in GA model was equal to that in HI model with half the admixture time.

Based on (2.4) and (2.6), we could observe that the variance of LACS in GA model was higher than in HI model if $T > 1$ (GA model assumes at least two generations since admixture). This was reasonable considering that admixed population in GA model contained both the long ancestral chromosomal segments that entered the admixed population recently and short ancestral chromosomal segments that entered much earlier. Besides, the mean and SD of LACS in HI model were identical (Fig. 2.3a) as they followed an exponential distribution, while the

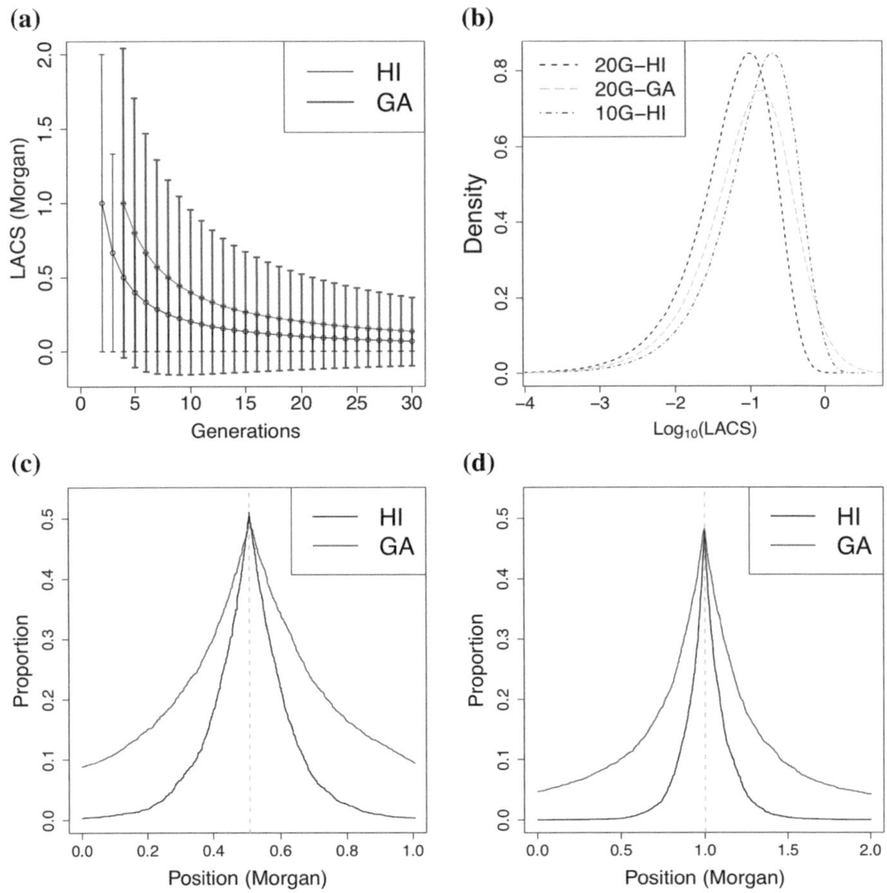

Fig. 2.3 Comparison of LACS distributions between HI model and GA model. **a** Comparison of mean and SD of LACS between HI model and GA model. Error bars and circles represent SD and mean, respectively. **b** Comparison of LACS distribution between HI model and GA model when generation is 10 or 20. **c** The change of genetic contribution that transmitted with the given locus along the chromosome in HI and GA models. **d** The change of genetic contribution that transmitted with the given locus along the chromosome in HI and GA models in a longer chromosome. *Green vertical dashed line* represents the reference locus. Reprint from Jin et al. [24]

SD of LACS in GA model was larger than the mean. Since variance of LACS distribution in GA model was larger than that in HI model with same generation (Fig. 2.3a), we conjectured the LACS distribution in GA model could be flatter than that in HI model, which was also supported by the observations (e.g., Fig. 2.3b). Although the mean LACS in GA model was the same as that in HI model with half the number of generations, there was a much higher proportion of long ancestral chromosomal segments in GA model compared to that in HI model (e.g., Fig. 2.3b).

After investigation of the overall pattern of LACS distribution in both HI and GA models, we further examined LACS distribution in a specific genomic region in the two classic models. A genetic locus was randomly selected from a gradually admixed population (admixed population under GA model) and a hybrid isolated population (admixed population under HI model). We found that the genetic contribution of loci transmitted with the given locus along the chromosome decreased quickly as the distance to the given locus increased in the hybrid isolated population while it decreased much slower in the gradually admixed population compared to that in hybrid isolated population (Fig. 2.3c, d). These results indicated that different admixture dynamics could have a strong influence on the pattern of local ancestral chromosomal segments. In particular, the change of genetic contribution of loci transmitted with the given locus in the gradually admixed population became slower as the distance to the locus increased. The genetic contribution of a given parental population to the admixed population was hardly reduced to 0 because some recent ancestral chromosomal segments from the given parental population were very long and even spread through the whole chromosome in the admixed population (Fig. 2.3d).

2.7 Influence of Distribution of LACS on Admixture Mapping

To compare the admixture mapping in HI and GA models, we used identical parameters for simulations by varying only the admixture model. Therefore, the main difference in the signature of association between hybrid isolated population and gradually admixed population should result from different admixture dynamics. In each model, we simulated 2,000 cases and 2,000 controls for admixture mapping, with genetic contribution of the given parental population to the admixed population $\theta = 20\,\%$, number of generations since the admixture $\lambda = 20$, and the increased risk of 2 for containing alleles from the given parental population at the susceptibility locus. Although the highest ancestral deviations at the susceptibility locus in both hybrid isolated population and gradually admixed population were identical (40 %), we found that the peak of association in the hybrid isolated population was narrower and sharper than that in gradually admixed population (Fig. 2.4), indicating that the identification of putative causal allele in hybrid isolated population could be more efficient than that in gradually admixed population. In contrast, the peak of signatures in gradually admixed population was wider than that in hybrid isolated population (Fig. 2.4), indicating that admixture mapping on a genome-wide scale in gradually admixed population required fewer markers than that in hybrid isolated population.

In the case-control designed admixture mapping, *p*-values were calculated by comparing the deviation of genetic contribution in cases with that in controls through phenotype association analysis [9, 23]. *P*-values of admixture mapping were determined from the ancestral deviation between cases and controls.

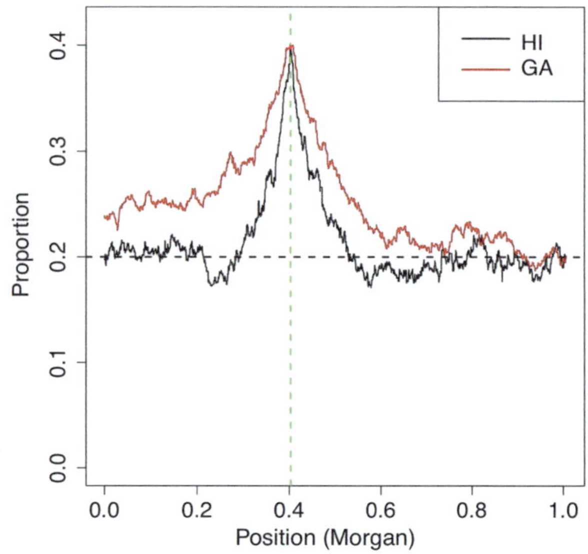

Fig. 2.4 Signatures of association in admixture mapping in HI and GA models. *Vertical dashed green line* represents the susceptibility locus. *Heretical dashed black line* represents the theoretical mean of genetic contribution of the given parental population. Reprint from Jin et al. [24]

Therefore, the *p*-value of susceptibility locus in GA model could be the same as that in HI because the highest ancestral deviations in both HI and GA models were identical. In contrast, the *p*-values in case-only designed admixture mapping were calculated based on the empirical distribution of LACS in cases [9, 23]. Since the distribution of signatures in GA model was wider than that in HI model, the *p*-value in GA model could be larger than that in HI model, which indicated that the signatures in HI were more likely to be significant compared to those in GA model. Therefore, we suggest case-control designed study rather than case-only designed study to improve the statistical power of admixture mapping in gradually admixed populations. In contrast, we suggest the case-only designed admixture mapping in hybrid isolated populations to reduce the cost.

References

1. HUGO Pan-Asian SNP Consortium et al. Mapping human genetic diversity in Asia. Science. 2009;326:1541–5.
2. Seldin MF, Pasaniuc B, Price AL. New approaches to disease mapping in admixed populations. Nat Rev Genet. 2011;12:523–8.
3. Verdu P, Rosenberg NA. A general mechanistic model for admixture histories of hybrid populations. Genetics. 2011;189:1413–26.
4. Wang S, et al. Geographic patterns of genome admixture in Latin American Mestizos. PLoS Genet. 2008;4:e1000037.
5. Tishkoff SA, et al. The genetic structure and history of Africans and African Americans. Science. 2009;324:1035–44.
6. Chakraborty R, Weiss KM. Admixture as a tool for finding linked genes and detecting that difference from allelic association between loci. Proc Natl Acad Sci U S A. 1988;85:9119–23.

7. McKeigue PM. Mapping genes underlying ethnic differences in disease risk by linkage disequilibrium in recently admixed populations. Am J Hum Genet. 1997;60:188–96.
8. McKeigue PM. Mapping genes that underlie ethnic differences in disease risk: methods for detecting linkage in admixed populations, by conditioning on parental admixture. Am J Hum Genet. 1998;63:241–51.
9. Montana G, Pritchard JK. Statistical tests for admixture mapping with case-control and cases-only data. Am J Hum Genet. 2004;75:771–89.
10. Pool JE, Nielsen R. Inference of historical changes in migration rate from the lengths of migrant tracts. Genetics. 2009;181:711–9.
11. Adams J, Ward RH. Admixture studies and the detection of selection. Science. 1973;180:1137–43.
12. Long JC. The genetic structure of admixed populations. Genetics. 1991;127:417–28.
13. Gravel S. Population genetics models of local ancestry. Genetics. 2012;191:607–19.
14. Price AL, et al. Sensitive detection of chromosomal segments of distinct ancestry in admixed populations. PLoS Genet. 2009;5:e1000519.
15. Tang H, Coram M, Wang P, Zhu X, Risch N. Reconstructing genetic ancestry blocks in admixed individuals. Am J Hum Genet. 2006;79:1–12.
16. Sankararaman S, Kimmel G, Halperin E, Jordan MI. On the inference of ancestries in admixed populations. Genome Res. 2008;18:668–75.
17. Zakharia F, et al. Characterizing the admixed African ancestry of African Americans. Genome Biol. 2009;10:R141.
18. Kidd JM, et al. Population genetic inference from personal genome data: impact of ancestry and admixture on human genomic variation. Am J Hum Genet. 2012;91:660–71.
19. Ewens WJ, Spielman RS. The transmission/disequilibrium test: history, subdivision, and admixture. Am J Hum Genet. 1995;57:455–64.
20. Pfaff CL, et al. Population structure in admixed populations: effect of admixture dynamics on the pattern of linkage disequilibrium. Am J Hum Genet. 2001;68:198–207.
21. Guo W, Fung WK. The admixture linkage disequilibrium and genetic linkage inference on the gradual admixture population. Yi Chuan Xue Bao. 2006;33:12–8.
22. Lilliefo, Hw. On Kolmogorov-Smirnov test for normality with mean and variance unknown. J Am Stat Assoc **62**, 399 (1967).
23. Patterson N, et al. Methods for high-density admixture mapping of disease genes. Am J Hum Genet. 2004;74:979–1000.
24. Jin W, Li R, Zhou Y, Xu S. Distribution of ancestral chromosomal segments in admixed genomes and its implications for inferring population history and admixture mapping. Eur J Hum Genet. 2014;22:930–7.

Chapter 3
Exploring Population Admixture Dynamics via Distribution of LACS

Abstract By comparing the distribution of LACS in admixed population with those of various hypothesized scenarios, we could find the scenario with distribution of LACS best fit the observed distribution. Based on 491,526 autosomal SNPs genotyped in thousands of African Americans, we found that a continuous gene flow model, in which African Americans continuously received gene flow from European parental population over about 14 generations, best explained the admixture dynamics of African Americans among several putative scenarios. On the contrary, the admixture dynamics of Mexicans was more likely to be explained by a gradual admixture model in which Mexicans continuously received gene flow from both European and Amerindian parental populations for about 24 generations.

Keywords Population admixture · Ancestral chromosomal segment · Admixture dynamics · African Americans · Mexicans

3.1 Different Admixture Dynamics, Distinct Distribution of LACS

Although we have inferred the theoretical distribution of LACS under HI model and GA model (Fig. 3.1a, b), there are much more complex admixture scenarios in reality. Here, we introduced another admixture model, continuous-gene-flow model (Fig. 3.1c) [1, 2], an extension of GA model. In all the three admixture models, m and $1 - m$ denote genetic contributions of the two parental populations pop1 and pop2 to the admixed population, respectively. T denotes the number of generations since admixture. In the CGF model, one parental population (genetic donor) continuously contributes a constant but reduced rate of gene flow (α) to

W. Jin, *Admixture Dynamics, Natural Selection and Diseases in Admixed Populations*, Springer Theses, DOI 10.1007/978-94-017-7408-6_3

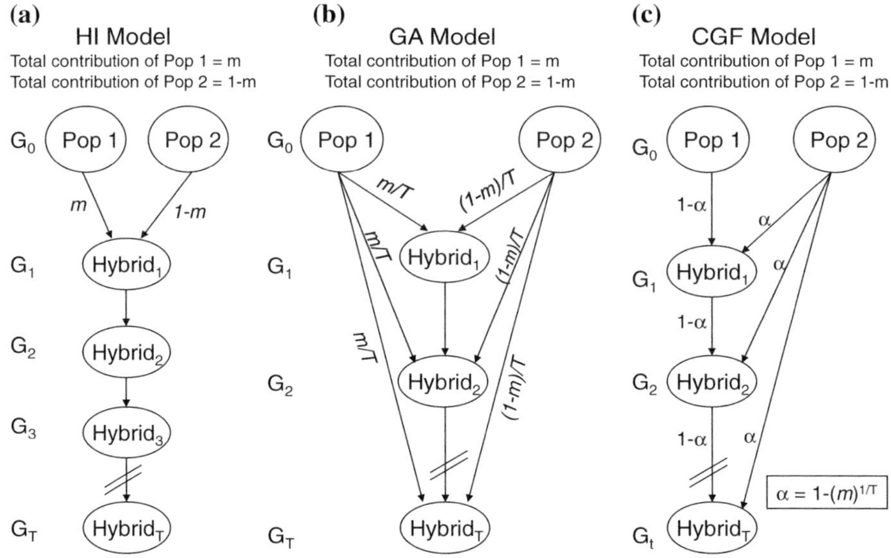

Fig. 3.1 Admixture models used to simulate the population admixture process. Hybrid isolation model **a** Graduate admixture model **b** and continuous-gene-flow model **c** In each model, the genetic contributions of pop1 and pop2 are m and $1 - m$, respectively. The admixed population experienced G_t generation, which ranges from 1 to T generation. Adapted from Jin et al. [23]

the admixed population per generation, so that the cumulative genetic contribution from both parental populations is equal to that under HI model and GA model. The gene flow that the admixed population receives from the genetic donor per generation is calculated using $\alpha = 1 - (m)^{1/T}$. Because the parental populations play different roles in the admixture process in CGF model, the parental population acting as genetic donor was referred to as CGFD, while the parental population acting as genetic recipient was referred to as CGFR. We investigated a scenario in which the genetic contribution of the parental population to the admixed population was 50 %. Numbers of generations since admixture for each model were set at 10, 20, 50, and 100. The basic information regarding the LACS indicated that different admixture processes and generations led to different distributions of LACS (Table 3.1). The distributions of LACS between different models differed significantly when the numbers of generations since admixture were the same ($P < 2.2 \times 10^{-16}$, two-sample K–S test). The distributions of LACS also differed significantly when the generations since admixture were different under the same model ($P < 2.2 \times 10^{-16}$, two-sample K–S test).

The 10-based logarithms of LACS were calculated because the distribution of LACS approximates a log-normal distribution. After this transformation, we could easily distinguish the distributions of LACS under different admixture models (Fig. 3.2). The distributions of LACS between different models were significantly different ($P < 2.2 \times 10^{-16}$, two-sample K–S test). To quantify the differences of

Table 3.1 Basic information on LACS when genetic contribution of the parental population to the admixed population is 50 %

Gen	Model	Mean (cM)	SD	SEM	P0	P25	P50	P75	P100
10	HI	17.838	17.372	0.017	0	5.24	12.57	24.97	232.32
10	CGF	31.79	38.937	0.052	0	7.19	18.22	40.33	278.08
10	CGFR	31.593	30.342	0.041	0	9.26	22.39	44.56	278.08
10	GA	34.742	38.95	0.055	0	8.69	21.7	46.08	278.08
20	HI	9.553	9.434	0.007	0	2.8	6.65	13.28	127.74
20	CGF	16.934	24.34	0.024	0	3.61	9.13	20.26	278.09
20	CGFR	16.939	16.739	0.016	0	4.89	11.79	23.61	208.86
20	GA	18.778	24.162	0.025	0	4.44	11.07	23.75	278.08
50	HI	3.969	3.926	0.002	0	1.17	2.77	5.49	55.29
50	CGF	7.159	11.962	0.008	0	1.48	3.73	8.3	278.08
50	CGFR	7.077	7.034	0.004	0	2.05	4.92	9.82	109.87
50	GA	7.863	11.534	0.008	0	1.82	4.53	9.7	278.08
100	HI	2.04	2.008	0.001	0	0.61	1.42	2.82	36.22
100	CGF	3.64	6.623	0.003	0	0.75	1.88	4.19	278.08
100	CGFR	3.629	3.597	0.002	0	1.06	2.53	5.03	65.86
100	GA	4.086	6.431	0.003	0	0.94	2.34	5.01	278.08

Gen Number of generations since admixture; *SD* Standard deviation; *SEM* Standard error of the mean; *P* represents percentiles

LACS distribution between the different models, earth mover's distance (EMD) [3, 4] was calculated. EMD is a measure of the dissimilarity between two probability distributions over a region. In each simulated scenario, the distributions of LACS under the GA model and the CGFR model were the most similar among the four models, whereas the HI model and the CGFR were the most dissimilar (Fig. 3.2). This observation was confirmed by EMDs between each pair of LACS distributions (Table 3.2). Within each model, the EMDs between the distributions of LACS increase as the numbers of generations since admixture increase (Table 3.2). The distribution of LACS was still significantly different when the shorter ancestral chromosomal segments were filtered out (Fig. 3.3), which indicated that admixture dynamics could be distinguished even if only long ancestral chromosomal segments were available.

3.2 Parental Populations of African Americans

African Americans are citizens or residents of the United States with partial recent Sub-Saharan African ancestry. The majority of them inherited their African ancestry from probably 500,000–650,000 Africans who were forcibly brought to the territories within the boundary of present United States during the Middle Passage [5, 6]. It is of importance to demonstrate the fine admixture dynamics of

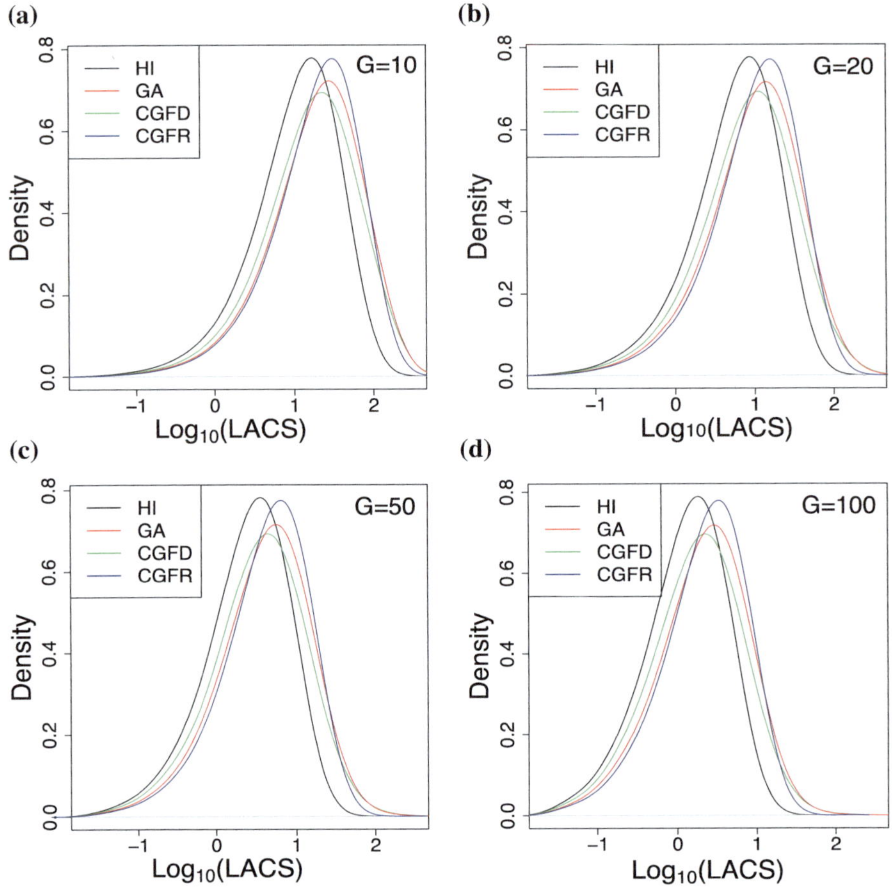

Fig. 3.2 Distributions of $\log_{10}$ (LACS) when genetic contribution of the parental population to the admixed population is 50 %. *G* number of generations since admixture. **a** Number of generations since admixture was set to 10. **b** Number of generations since admixture was set to 20. **c** Number of generations since admixture was set to 50. **d** Number of generations since admixture was set to 100. Adapted from Jin et al. [23]

African Americans because of its strong implication in admixture mappings for disease gene discoveries. The populations from West Africa and Europe undoubtedly contributed to the current gene pool of African Americans, therefore, they were considered as the parental populations of African Americans in this study. In fact, the selection of proper populations that contributed to African Americans is practically difficult because the extant populations in West Africa and Europe may not necessarily represent those of 300 years ago, given the possible influence of genetic drift, selection, and demographic history. The genetic contribution of multiple populations in West Africa and Europe further complicated the endeavor.

Table 3.2 EMDs between each pair-wise distributions of $Log_{10}LACS$ when genetic contribution of the parental population is 50 %

	10G_HI	10G_CGF	10G_CGFR	10G_GA	20G_HI	20G_CGF	20G_CGFR	20G_GA	50G_HI	50G_CGF	50G_CGFR	50G_GA	100G_HI	100G_CGF	100G_CGFR
10G_CGF	0.184	0	–	–	–	–	–	–	–	–	–	–	–	–	–
10G_CGFR	0.249	0.082	0	–	–	–	–	–	–	–	–	–	–	–	–
10G_GA	0.247	0.066	0.03	0	–	–	–	–	–	–	–	–	–	–	–
20G_HI	0.272	0.456	0.518	0.52	0	–	–	–	–	–	–	–	–	–	–
20G_CGF	0.126	0.295	0.359	0.36	0.161	0	–	–	–	–	–	–	–	–	–
20G_CGFR	0.026	0.21	0.273	0.275	0.245	0.103	0	–	–	–	–	–	–	–	–
20G_GA	0.058	0.221	0.285	0.286	0.234	0.074	0.04	0	–	–	–	–	–	–	–
50G_HI	0.648	0.832	0.894	0.896	0.376	0.537	0.621	0.61	0	–	–	–	–	–	–
50G_CGF	0.494	0.678	0.74	0.742	0.227	0.382	0.468	0.456	0.154	0	–	–	–	–	–
50G_CGFR	0.402	0.586	0.649	0.651	0.13	0.291	0.376	0.365	0.246	0.109	0	–	–	–	–
50G_GA	0.42	0.604	0.667	0.669	0.153	0.309	0.394	0.383	0.227	0.074	0.044	0	–	–	–
100G_HI	0.932	1.116	1.179	1.18	0.66	0.821	0.906	0.894	0.284	0.438	0.53	0.511	0	–	–
100G_CGF	0.785	0.969	1.032	1.034	0.514	0.674	0.7590	0.748	0.15	0.292	0.385	0.365	0.146	0	–
100G_CGFR	0.688	0.872	0.935	0.937	0.417	0.578	0.662	0.651	0.041	0.195	0.286	0.268	0.243	0.115	0
100G_GA	0.704	0.888	0.951	0.953	0.433	0.593	0.678	0.667	0.072	0.211	0.303	0.284	0.227	0.082	0.044

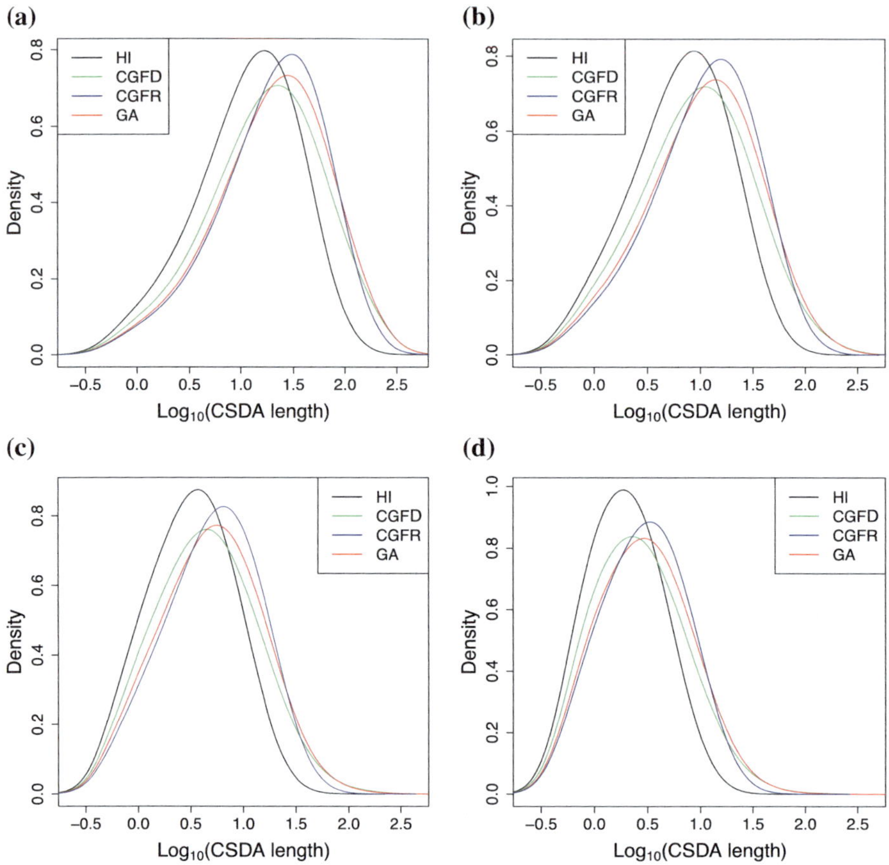

Fig. 3.3 Distributions of $\log_{10}$ (LACS) for LACS >0.5 cM when the genetic contribution of the parental population to the admixed population is 50 %. **a** Number of generations since admixture was set at 10. **b** Number of generations since admixture was set at 20. **c** Number of generations since admixture was set at 50. **d** Number of generations since admixture was set at 100

In particular, an exploratory analysis including Human Genome Diversity Project (HGDP) populations [7] also suggested that YRI (Yoruba in Ibadan, Nigeria) and CEU (Utah residents with northern and western European ancestry from the CEPH collection) are better choice for parental populations of African Americans than the others on which the data are available (Fig. 3.4). The 1,890 African American samples were studied together with 3,320 samples from 19 putative parental populations. In the PCA plot, populations from the same continents were clustered together, which reflected genetic diversity of populations from the same continent is rather lower compared with that of inter-continent populations. African Americans lie on the direct line between European (except Russian and Adygei, which may have some components from Asia) and African,

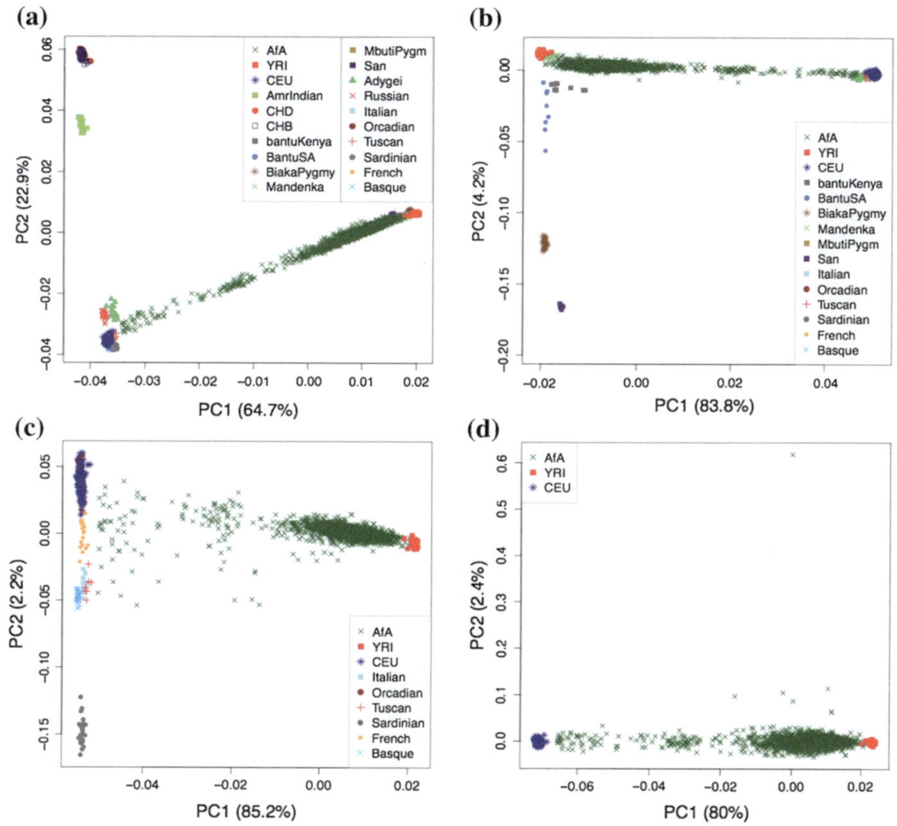

Fig. 3.4 Principal component analysis of African American samples and samples from the putative parental populations. *AfA* represents African Americans. The % eigenvalue is the percentage of the total variance in the top ten PCs. **a** 1,890 African Americans and 3,320 samples from 19 other populations represent global-wide populations. **b** 1,890 African Americans and all possible European or African parental populations. **c** 1,890 African Americans, YRI, and all possible European parental populations. **d** 1,890 African Americans and two putative parental populations (YRI and CEU)

which reflected their varying levels of admixture in African Americans between the two ancestral groups (Fig. 3.4a). When samples from Asia and America (as well as Russian and Adygei) were removed, African Americans almost distributed only between West African and European, which reflected the West African, especially YRI, contributing to most of the African components of ancestry (Fig. 3.4b). Substructure in European populations showed up when only YRI, African Americans, and European populations were analyzed. All European populations except Sardinian greatly contributed to European components of African Americans. However, CEU contributed to most of the European components of African American as only a few samples were beyond the line between YRI and

Fig. 3.5 FRAPPE analysis of YRI, African Americans, and CEU when $K = 2$. Individuals are represented as thin vertical lines partitioned into K segments corresponding to the inferred membership of the genetic clusters, which are indicated by the colors

CEU (Fig. 3.4c). Then, we performed PCA on samples from YRI, CEU, and African American. The PC1 explained 80 % of the total variance of top ten PCs while PC2 explained only 2.4 %, which reflect that most of the variances were caused by the genetic differentiation between West African and European ancestry, especially like YRI and CEU (Fig. 3.4d).

We chose YRI and CEU to represent those who contributed to the formation of African Americans, also largely due to the availability of phased genetic data of the two populations. FRAPPE [8] was applied to the genome-wide high-density SNPs data (by taking $K = 2$), and the estimated European contribution to African Americans was 21.65 % at population level (Fig. 3.5). These estimations were essentially consistent with those in previous studies [9–11], although different datasets were used. Based on the maximum likelihood estimated from HAPMIX, the number of generations since admixture (λ) was estimated to be 7, which could be considered as an average value based on HI model. The estimation was similar to those reported in other studies [9, 12–15]. However, it has been >300 years (15 generations assuming 20 years per generation) since the Atlantic slave trade in Seventeenth and Eighteenth century that brought most of the African ancestors of current African Americans to the New World. Therefore, the number of generations since admixture estimated by genetic data was not consistent with historical records.

3.3 Admixture Dynamics of African Americans

In order to resolve the contradiction between historical records and the number of generations estimated by genetic data, we simulated a serial of African American populations based on the four admixture models. In these simulations, European and African parental populations contributing to African American were set at 21.65 and 78.35 % according to the aforementioned observation, respectively. The numbers of generations for the simulated African American populations were set from 10 to 17, stepping by one generation. The empirical distributions of LACS for both African and European ancestral components were used HAPMIX output. We compared the empirical distributions of LACS to those of simulated data under the four models: HI model, GA model, CGF1 model (European population serving as CGFD and African as CGFR), and CGF2 model (European population serving

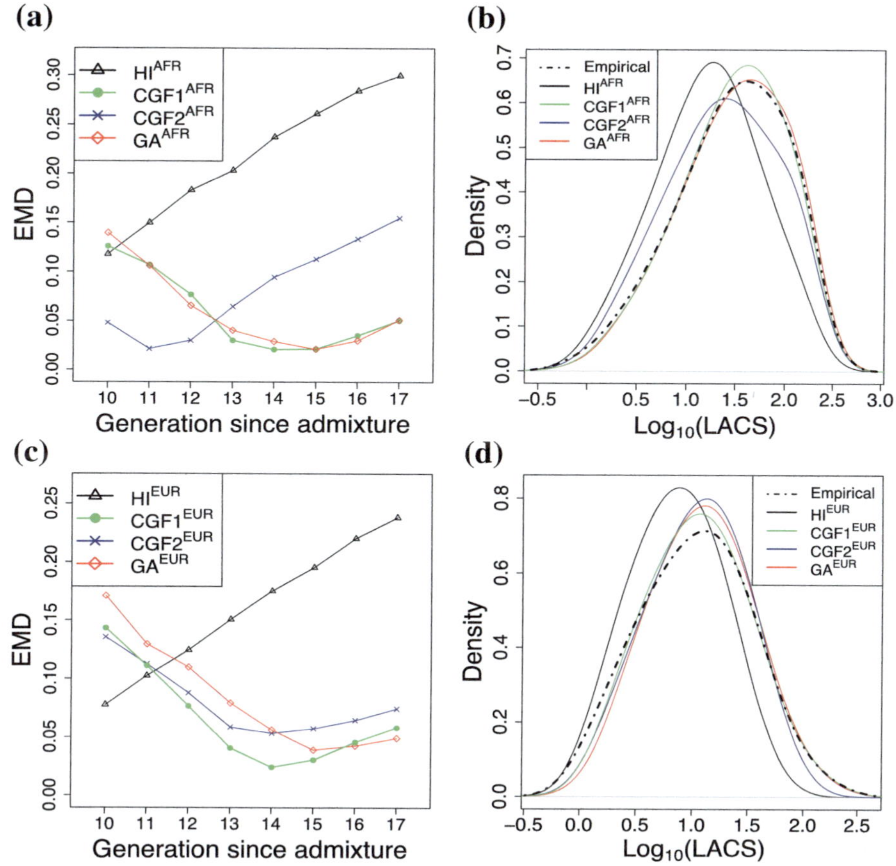

Fig. 3.6 Admixture dynamics of African Americans. For CGF model, the case in which Europeans who continually served as genetic donor were named CGF1 model, while Africans as genetic donor were named CGF2 model. To find the model that fit the empirical distribution best, earth mover's distance (EMD) between empirical distribution and that of each model was calculated. The model showing the lowest EMD with the empirical data was considered best fit. **a** Distribution of EMDs for African ancestral components between empirical data and each model. **b** Empirical distribution of LACS for African ancestral components and simulated distributions when the number of generations was set to 14. **c** Distribution of EMDs for European ancestral components between empirical data and each model. **d** Empirical distribution of LACS for European ancestral components and the simulated distributions when the number of generations was set to 14. Adapted from Jin et al. [23]

as CGFR and African as CGFD). To identify the model with distribution of LACS that fit the empirical distribution best, the EMD between empirical distribution and that of each model was calculated. The admixture model showing the lowest EMD with the empirical data was considered fit best. The lowest EMDs for both African and European ancestral components (EMD = 0.0204 and 0.0239, respectively) were observed under the 14-generation CGF1 model (Fig. 3.6a, c), in which gene flow ($\alpha = 0.017$) from European parental population continuously contributed to the African Americans.

For both African and European ancestries, the EMDs between empirical distributions and those of the HI model increased as the number of generations increased (Fig. 3.6a, c). Even the distributions of LACS in the 10-generation HI model were deficient in long ancestral chromosomal segments compared with empirical distributions for both European and African ancestral components (data not show). Therefore, the HI model can be excluded because the history of the African American population goes back to >10 generations (>200 years) based on the history records. In CGF2 model, gene flow from African parental population continuously contributed to the African American, the lowest EMD for African ancestral components between the empirical distribution and simulated distributions was obtained when generation since admixture was set at 11 (Fig. 3.6a). This observation was not consistent with that of European ancestral components, which shift to the lowest EMD at 14 generations (Fig. 3.6a, c). Furthermore, the lowest EMD for European ancestral components between empirical and simulated distributions in the CGF2 model was higher than that under either the CGF1 or the GA model (Fig. 3.6c). Therefore, the CGF2 model could not hold when both African and European ancestries were considered. The distribution of EMD between GA model and empirical data was similar to that of CGF1 model when African ancestral components were investigated (Fig. 3.6a), However, the lowest EMD for European ancestral components under GA model always higher than that under CGF1 model.

Although the actual population admixture dynamics of African Americans could be much more complex than our simulations, the CGF1 model with 14 generations was found to be reasonably representative, capturing the major feature of the population admixture. Direct comparison of the empirical distribution of LACS with that of the simulated admixture populations with 14 generations also supported the CGF1 model (Fig. 3.6b, d), although the empirical distribution of European ancestral components was slightly flatter than the simulated distribution, which possibly resulted from non-random mating. In fact, the admixture model fits the social reality of African Americans in United States quite well. On one hand, the African immigrants to the United States have been rare in recent 200 years since the ban of the Atlantic slave trade, there is no obvious recent gene flow from Africans. On the other hand, it is well known that inter-marriage or inter-breeding between African Americans and European Americans existed in every period since the African immigrants arrived in the New World, although it has not happened on a large scale. Therefore, gene flow from Europe Americans continuously contributed to gene pool of the African Americans because children with one European parent and one African American parent were generally regarded as African Americans. Since the gene flow from the European population is expected to continually contribute to the African American gene pool, it is very likely that the overall genetic contribution of European parental population to African Americans will continuously increase in the future.

3.4 Substructure in African American Admixture

The ancestry proportion of each African American could provide information about the admixture dynamics. However, the distributions of individual ancestry proportions for African Americans fit none of the four models perfectly (Fig. 3.7a). This may have been due to the small sample size, sampling error, or substructure within the African American population. By carefully examining the distribution of individual ancestry proportions, we found that a small fraction of African Americans had a much higher proportion of European ancestry (or with very little African ancestry) than those simulated (Fig. 3.7a), indicating substructures of African American population in terms of ancestry proportion. The observation might result from individuals from particular lineages having been continuously inter-married with people of European ancestry or with dominant European ancestry but not people of general African American, generation by generation. We found that individuals with extremely high proportion of African ancestry or European ancestry tended to have more estimated generations since admixture (Fig. 3.7b), which might suggest that the individuals with only little European or African ancestral components tended to inherit it from much earlier admixture events, while those with roughly even ancestral contributions were more likely to have descended from more recent admixture events.

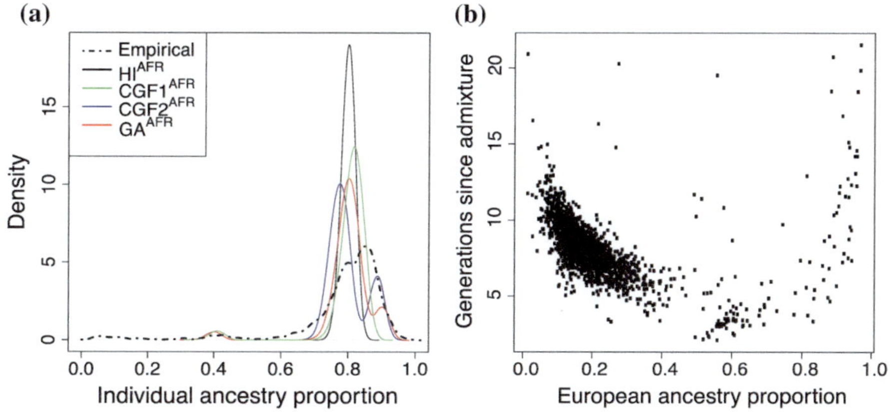

Fig. 3.7 Substructure showed by the individual ancestry proportion. **a** Empirical distribution of individual ancestry proportion for African ancestry components and simulated distribution when the number of generations was set at 14. **b** Relationship between ancestry proportion and estimated number of generations since admixture for each African-American individual. Each black points correspond to one of the 1,890 African Americans who studied here

3.5 Admixture Dynamics of Mexicans

Mexicans, including both Mexican-Americans and Mexican-Mestizos, are the second most widely used population in admixture mapping. Based on simplified model without considering admixture dynamics, the generation since admixture of Mexican-Americans or Mexican-Mestizos has been estimated to be ≤ 15 by average LACS in other studies [16–18]. However, the admixture history of Mexicans is much longer considering that Europeans first colonized the New World more than 500 years ago (>25 generations assuming 20 years per generation). In order to simply the admixture analysis, we filtered out the Mexican samples with obvious African ancestral components. Overall, 423 Mexican individuals with negligible recent ancestry other than Amerindian and European were investigated (Fig. 3.8). In the PC plot, four continental populations (YRI, CHB, CEU, and AMI) were located at the corner of ladder-shaped plot, while all these 423 Mexican individuals projected between CEU and Amerindian clusters (Fig. 3.8). The genetic contribution of European ancestry to the 423 Mexican samples was estimated to be 49.2 % using PCA.

We simulated a serial of admixed populations by setting the genetic contribution of European to the admixed populations at 49.2 % and 15–25 generations since admixture, stepped by one generation. The empirical distributions of LACS for both Amerindian and European ancestries were obtained by merging ancestral chromosomal segments from Mexican-Americans and Mexican-Mestizos that had been inferred using HAPMIX. The EMDs between the HI model and empirical distribution for both Amerindian and European ancestral components increased as the number of generations since admixture increased (Fig. 3.9a, c), and distributions under HI model at 15 generations still lacked long ancestral chromosomal

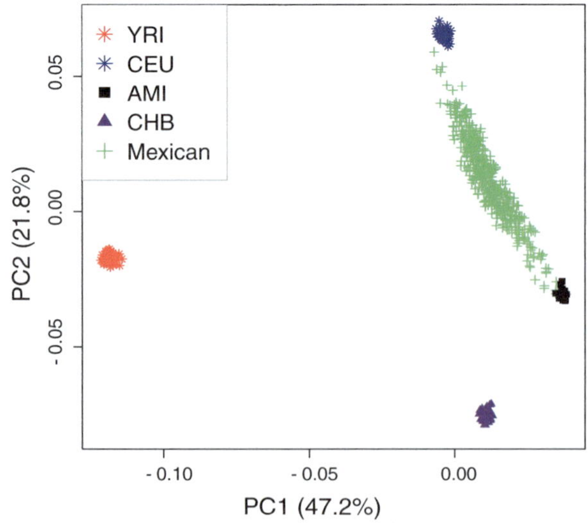

Fig. 3.8 Principal component analysis of Mexican samples and samples from the putative parental populations. The number in parentheses is the percentage of the total variance in the top ten PCs. It is shown that the filtered 423 Mexicans dispersed between CEU and AMI. AMI represents Amerindian

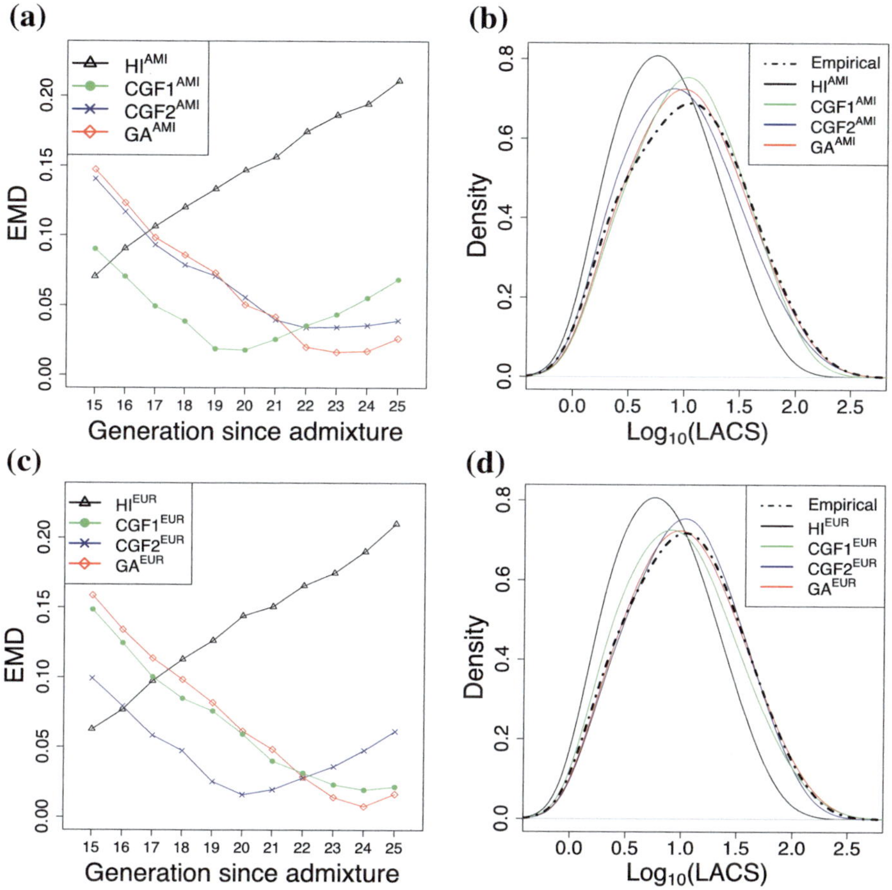

Fig. 3.9 Admixture dynamics of Mexicans. The model showing the lowest EMD with the empirical data was considered as the best fit. The GA model, in which both European and Amerindian populations continuously contributed to the Mexican gene pool over about 24 generations, fit the empirical data best. **a** Distribution of EMDs for Amerindian ancestral components between empirical data and each model. **b** Empirical distribution of LACS for Amerindian ancestral components and the simulated distributions when the number of generations was set to 24. **c** Distribution of EMDs for European ancestral components between empirical data and each model. **d** Empirical distribution of LACS for European ancestral components and the simulated distributions when the number of generations was set to 24. Adapted from Jin et al. [23]

segments (data not shown). For both the CGF1 model (European population serving as CGFD and Amerindian as CGFR) and the CGF2 model (European population serving as CGFR and Amerindian as CGFD), the lowest EMDs for Amerindian and European ancestral components were inconsistent. Especially, both the lowest EMDs generated by these two models were still higher than that generated using the GA model. This indicated that the GA model fit the empirical data best among the four models. The EMDs between empirical distributions and distributions under GA model for both Amerindian and European ancestral components reached the lowest value (EMD = 0.0163 and 0.0076, respectively) at 23 and 24 generations, respectively (Fig. 3.9a, c). In short, the GA model at 24 generations fit the empirical data best out of all these simulations, as indicated by the distribution of EMDs. Direct comparisons also support that empirical distribution of LACS essentially fit the 24-generation GA model (Fig. 3.9b, d).

3.6 Genetics Difference Between Mexican Americans and Mestizos

Considering that both pure Amerindian and pure European migrants have coexisted in Mexico, the GA model is intuitively much more reasonable than the others. When we considered the Mexican-Americans and Mexican-Mestizos separately, we found the genetic contribution of European ancestry to Mexican-Americans to be 53.9 %, which was significantly higher than that of 268 Mexican-Mestizos (46.7 %, $P = 0.0018$, t test) (Fig. 3.10). Further analysis showed that the distribution of LACS of Amerindian ancestral components in Mexican-Americans was essentially identical to that of Mexican-Mestizos (Fig. 3.11a). However, the LACS of European ancestry in Mexican-Americans were much longer than those present in Mexican-Mestizos (Fig. 3.11b), which suggested the recent gene flow from European to Mexican-American populations. In other words, the fact that European populations have contributed more to Mexican-Americans than to Mexican-Mestizos was likely to have resulted from recent inter-marriage between European-Americans and Mexican immigrants in the U.S.

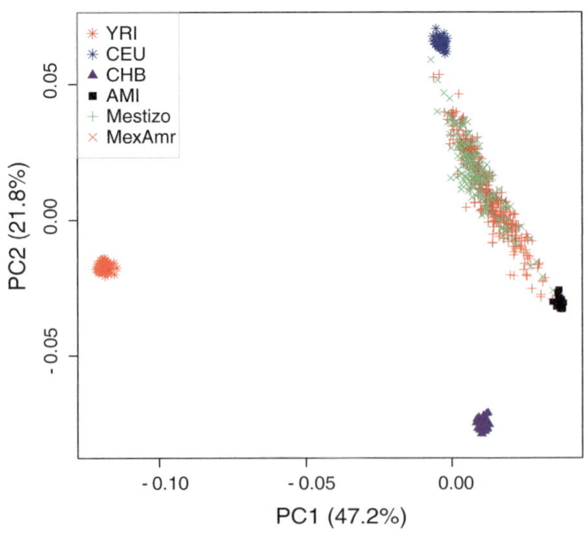

Fig. 3.10 Principal component analysis of Mexican-American and Mestizo samples and samples from putative parental populations. *MexAmr* represents Mexican-American. The proportion in parentheses is the percentage of the total variance in the top ten PCs

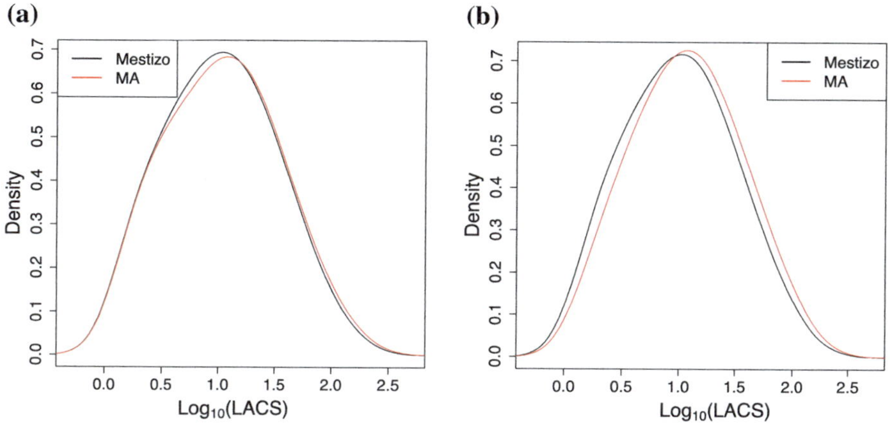

Fig. 3.11 Comparisons of admixture dynamics between Mexican-American and Mestizo. *MA* represents Mexican American. **a** Distribution of LACS for Amerindian ancestral components in Mexican-Americans and Mestizos. **b** Distribution of LACS length for European ancestral components in Mexican-Americans and Mestizos

3.7 Implication for Future Study

Population admixture is influenced by various social, cultural, economic, and geographical factors, which lead to much complex admixture processes. Therefore, we did not expect that the actual admixture process could be fully explained by any single simplified model. However, in practice, to facilitate the evolutionary and medical studies that rely on the knowledge of admixture dynamics, we suggest the primary pattern should be captured. In this study, by comparing the empirical distribution of LACS with those under the simplified models, we determined the primary admixture model of two most important admixed populations and provided new insights into their admixture dynamics. In detail, we showed that two-way admixture dynamics of African Americans best fit the 14-generation CGF model among all the four possible scenarios, in which European ancestry continually contributed to the African American gene pool. Second, we showed that Mexican data fitted the 24-generation GA model best and recent gene flows from European population might have contributed to the Mexican-American gene pool. Therefore, the distribution of LACS has been shown to be powerful in distinguishing different admixture models in various scenarios.

New World admixed populations such as African Americans are widely used in disease gene discovery through admixture mapping. The effects of admixture dynamics on the pattern of linkage disequilibrium have been analyzed in various studies [19–22]. However, most previous studies simulated the African American population simply using an HI model, and assumed an admixture time of only 6–8 generations, which were the average values indicated by genetic data [13, 14]. The real statistical power in admixture mapping may have been significantly affected in those studies because the admixture dynamics of African Americans, as shown in this study, are more likely to fit the 14-generation CGF model in which gene flow from European parental population to African Americans was $\alpha = 0.017$ per generation. We suggest that future studies should simulate African Americans based on the admixture dynamics demonstrated in this study for accurately evaluating the statistics power of admixture mapping. To our knowledge, this is the first attempt to explore population admixture dynamics using genome-wide distribution of LACS. The much complex admixture dynamics could be explored when the accuracy of ancestral chromosomal segments' inference would be significantly improved in the future.

References

1. Pfaff CL, et al. Population structure in admixed populations: effect of admixture dynamics on the pattern of linkage disequilibrium. Am J Hum Genet. 2001;68:198–207.
2. Long JC. The genetic structure of admixed populations. Genetics. 1991;127:417–28.
3. Hitchcock FL. The distribution of a product from several sources to numerous localities. J Math Phys. 1941;20:224–30.
4. Rubner Y, Tomasi C, Guibas LJ. The earth mover's distance as a metric for image retrieval. Int J Comput Vision. 2000;40:99–121.
5. Zakharia F, et al. Characterizing the admixed African ancestry of African Americans. Genome Biol. 2009;10:R141.
6. Thomas H. The slave trade: the story of the Atlantic slave trade. Simon Schuster. 1999;1440–1870.
7. Li JZ, et al. Worldwide human relationships inferred from genome-wide patterns of variation. Science. 2008;319:1100–4.
8. Tang H, Peng J, Wang P, Risch NJ. Estimation of individual admixture: analytical and study design considerations. Genet Epidemiol. 2005;28:289–301.
9. Smith MW, et al. A high-density admixture map for disease gene discovery in african americans. Am J Hum Genet. 2004;74:1001–13.
10. Xu S, et al. Dissecting linkage disequilibrium in african-american genomes: roles of markers and individuals. Mol Biol Evol. 2007;24:2049–58.
11. Bryc K, et al. Genome-wide patterns of population structure and admixture in West Africans and African Americans. Proc Natl Acad Sci USA. 2010;107:786–91.
12. Price AL, et al. Sensitive detection of chromosomal segments of distinct ancestry in admixed populations. PLoS Genet. 2009;5:e1000519.
13. Tian C, et al. A genomewide single-nucleotide-polymorphism panel with high ancestry information for African American admixture mapping. Am J Hum Genet. 2006;79:640–9.
14. Seldin MF, et al. Putative ancestral origins of chromosomal segments in individual african americans: implications for admixture mapping. Genome Res. 2004;14:1076–84.
15. Smith MW, et al. A high-density admixture map for disease gene discovery in african americans. Am J Hum Genet. 2004;74:1001–13.
16. Wang S, et al. Geographic patterns of genome admixture in Latin American Mestizos. PLoS Genet. 2008;4:e1000037.
17. Tian C, et al. A genomewide single-nucleotide-polymorphism panel for Mexican American admixture mapping. Am J Hum Genet. 2007;80:1014–23.
18. Price AL, et al. A genomewide admixture map for Latino populations. Am J Hum Genet. 2007;80:1024–36.

19. Smith MW, O'Brien SJ. Mapping by admixture linkage disequilibrium: advances, limitations and guidelines. Nat Rev Genet. 2005;6:623–32.
20. Pfaff CL, et al. Population structure in admixed populations: effect of admixture dynamics on the pattern of linkage disequilibrium. Am J Hum Genet. 2001;68:198–207.
21. Pfaff CL, Kittles RA, Shriver MD. Adjusting for population structure in admixed populations. Genet Epidemiol. 2002;22:196–201.
22. Guo W, Fung WK. The admixture linkage disequilibrium and genetic linkage inference on the gradual admixture population. Yi Chuan Xue Bao. 2006;33:12–8.
23. Jin W, Wang S, Wang H, Jin L, Xu S. Exploring population admixture dynamics via empirical and simulated genome-wide distribution of ancestral chromosomal segments. Am J Hum Genet. 2012;91:849–62.

Chapter 4
Genome-Wide Search for Signatures of Natural Selection in African Americans

Abstract The several genomic regions showing excess of African or European ancestry could be the footprints of natural selection in the post-admixture era. In order to detect the natural selection since the African or European ancestry left for the New World, we reconstructed an artificial African population using the inferred African ancestral chromosomal segments and compared it with indigenous African populations. Interestingly, many selection-candidate genes identified by the later approach were associated with African American specific high-risk diseases such as prostate cancer and hypertension, suggesting these disease-related genes might have played an important role in African Americans adapting to the new environment. *CD36* and *HBB*, whose mutations confer a degree of protection against malaria, were also located in those highly differentiated regions between the artificial African population and indigenous African population. Further analysis showed frequencies of alleles protecting against malaria were lower in artificial African population than in indigenous African population, which consists with the relaxed selection pressure of malaria in U.S.

Keywords Natural selection · African Americans · Population admixture · Ingenuity pathway analysis (IPA) · Single nucleotide polymorphism (SNP)

4.1 Potential Natural Selection Events in African Americans

Based on modern population genetics, the vast majority of the genetic variations in human genome evolve neutrally. However, the variations in some loci have been shaped by natural selection in adapting population to its local environment [1, 2]. Investigation on the signatures of natural selection have led to the discovery of

W. Jin, *Admixture Dynamics, Natural Selection and Diseases in Admixed Populations*, Springer Theses, DOI 10.1007/978-94-017-7408-6_4

genes that played an important role in adaptation population to local pathogens, climate, diet, and other environmental challenges. These discoveries have greatly enriched our knowledge on our human origins and evolutionary history, led the identification of genes with important biological functions, and further demonstrated the genetic basis of some human diseases [3–7]. The recent availability of high-density SNPs data has provided the essential resources for genome-wide search for signatures of natural selection, especially for ethnically well-defined populations without obvious admixture [3, 4, 8, 9]. Although several studies have investigated natural selection in recently admixed populations [10–12], no study has so far investigated the locus-specific population differentiation between the ancestral components of the admixed population and its parental populations, which might reflect the natural selection since the two split. African American is a well-studied admixed population and provides a unique opportunity for such analysis.

African Americans are citizens or residents of the United States with any fraction of recent Sub-Saharan African ancestry. The majority of them inherited their African ancestry from probably 500,000–650,000 native Africans who were forcibly brought to the territories within the boundary of present United States during the Middle Passage [13, 14]. There is much high mortality during the Middle Passage due to the inappropriate conditions on board, and during the first several years after their arrival in the New World as a result of exposure to foreign pathogens and poor living conditions. Although the exact amount of life lost during the period remains a mystery, it may equal or exceed the amount actually enslaved in the plantations [15]. In this way, we could attribute the high mortality of the African Americans in the slavery era to great environmental challenges such as foreign pathogens. These persistent selection forces might make the frequencies of beneficial alleles to increase continuously, which could lead to higher population differentiation between the African ancestral components in African Americans and the indigenous Africans at these loci, in contrast to the great majority of loci that evolved neutrally.

Meanwhile, before African and European immigrants left for the New World, their ancestries had evolved independently in the distinct environments for tens of thousands of years [11, 12]. It is a great challenge for the two immigrant populations and the population of their admixture to adapt to the completely new environment in the New World. When an allele from the particular parental population shows advantage over its counterpart in another parental population, the frequency of the allele will likely increase in the admixed population due to the survival advantage, which will lead to ancestral components carrying this advantage allele increase generation generationally in the admixed population. Therefore, the genomic regions in African Americans showing an excess of a particular ancestral component are attributed to natural selection after population admixture [11, 12]. In fact, various studies investigated the deviations of genetic contributions for signatures of natural election in African Americans based on the limited number of loci that were available at that time [16–18]. Based on approximately 500 K SNPs genotyped in 365 African Americans, Bryc et al. [10] identified three autosomal regions showing excessive or reduced African ancestral components as natural selection candidates.

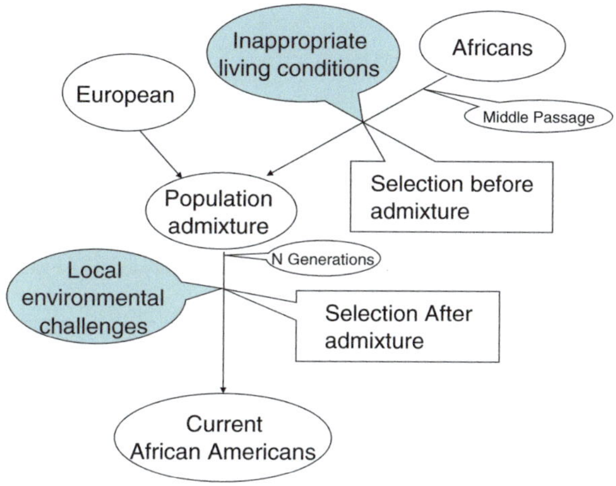

Fig. 4.1 Schematic of pre-admixture selection and post-admixture selection. The natural selection category was based on the changing environment challenge at different period

According to African Americans' development history, natural selection in African Americans could be classified into natural selection before population admixture and natural selection after population admixture (Fig. 4.1), both of which associated with distinct environment challenges. We could detect the signatures of natural selection that were more likely to occur after admixture by examining the genome-wide distribution of ancestry in African Americans. However, it is not easy to detect the signatures of nature selection before population admixture because the population before population admixture but experienced the pre-admixture natural selection is no longer available. Then we developed a new strategy by reconstructing an artificial African population (AAF) from inferred African ancestral components in African Americans and comparing it with indigenous African populations (Fig. 4.2), which reflect natural selection since the African ancestors of African Americans left Africa (including both pre- and post-admixture).

4.2 Identification of Genomic Regions with Biased Ancestry

The aforementioned 1,890 African Americans, in which European contributed 21.65 %, were used for investigating the signatures of natural selection. We have demonstrated that the African American admixture dynamics fit a 14-generation CGF model, in which gene flow ($\alpha = 0.017$) from European parental population continuously contributed to the African Americans. The locus-specific genetic contribution of European parental population to African Americans across the genome was estimated to be 21.68 ± 0.75 % (mean $\pm$ SD). The standard deviation (SD) of the locus-specific genetic contributions is lower than those of any previous studies on the New World admixed populations [10–12, 16–18], which was not unexpected since we used a much larger sample size. The six genomic

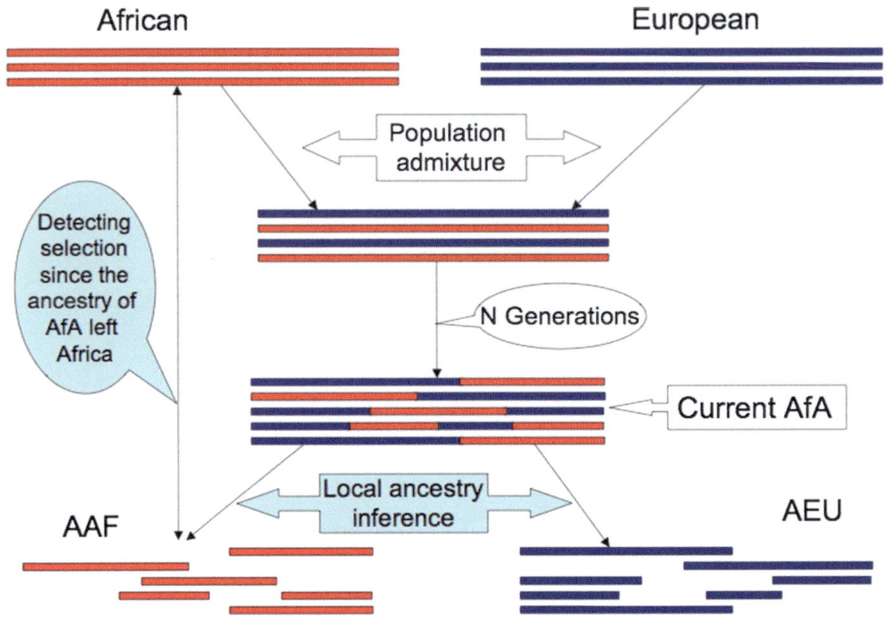

Fig. 4.2 Schematic of the strategy for detecting natural selection by comparing the AAF with indigenous African populations. "AfA" and "AAF" represent African Americans and artificial African population, respectively

regions showing strong deviation of European ancestral components (3 SDs above or below the genome-wide average) were treated as candidates of natural selection (Fig. 4.3), among which were four regions with excessive European influence (2p22, 3q13, 6q26, 16q21) and two regions with excessive African influence

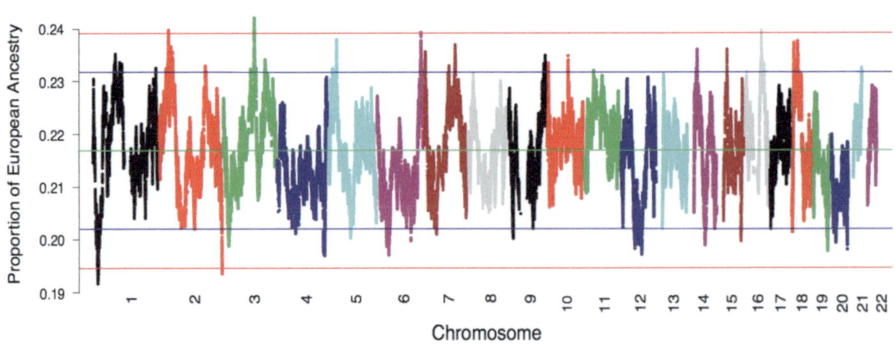

Fig. 4.3 Genome-wide Distribution of European Ancestral Contributions. Mean European ancestral contribution across 1,890 African American individuals at each SNP. *Green line* is the estimated genome-wide mean European ancestral contribution (21.68 %). *Blue bands* indicate +2 and –2 SDs from the mean ancestral contribution and *red bands* indicate +3 and –3 SDs from the mean ancestral contribution. Reprint from Jin et al. [55]

Table 4.1 Regions showing excess of European or African ancestry

Regions	Position	Excess ancestry	Size (bp)	SNPs	Highest deviation	Genes	Pathways	Related diseases
1p36	chr1:17409539..21604321	African	4,194,783	489	0.0253	*AKR7A2**, *IGSF21*, *DDOST**, HTR6 etc.	Diabetes pathways, signaling by GPCR, metabolism of amino acids	Diabetes, pancreatic cancer
2q37	chr2:241750403..242568618	African	818,216	16	0.0231	*SEPT2**, *HDLBP**, *PDCD*1*, *FARP2* etc.	Signaling in immune system, Axon guidance, metabolism of nucleotides	Bladder cancer, lung cancer, coronary atherosclerosis
2p22	chr2:37451925..37508581	European	56,657	9	0.0230	*QPCT*, (*EIF2AK2** 222 kb)	Influenza infection	Influenza infection
3q13	chr3:116930811..118313302	European	1,382,492	216	0.0253	*LSAMP**	Homophilic adhesion	Osteosarcoma
6q26	chr6:163653158..163653428	European	271	2	0.0225	*PACRG**	Mediate proteasomal degradation	Juvenile Parkinson's disease
16q21	chr16:61214438..61242497	European	28,060	9	0.0229	NA	NA	NA

NA: not available. * Denotes genes associated with diseases. Genes in parentheses are strong candidates out of the chromosome location but closest

(1p36, 2q37). The genetic contribution of each of the six regions was significantly different from the genome-wide ancestral contributions ($P < 2.2 \times 10^{-16}$, t test). The detailed annotations of the six candidate regions are presented in Table 4.1. Most genomic regions showing ancestry deviation can be replicated by LAMP-ANC, although it has lower accuracy than HAPMIX based on our simulated data.

Examination of SNPs in the six genomic regions showed neither significant deviation from Hardy–Weinberg expectation nor unusual fraction of missing data, suggesting that the genotyping quality is unlikely the cause of the bias. The six regions did not overlap with previously reported long-ranged linkage disequilibrium blocks which may confound genome-wide scans for selection signals in admixed populations [19]. Although the three regions showing biased ancestry in Bryc et al. [10] were in moderate excess of African or European ancestral components in this study, the estimated locus-specific ancestral contributions across the genome are generally consistent.

The genomic region showing the highest excessive African ancestral components is located in 1p36, in which have two strong signals near each other. *IGSF21* and *AKR7A2* are located next to the SNPs showing strongest signals, respectively. *IGSF21* belongs to immunoglobin superfamily, while *AKR7A2* is involved in the detoxification of aldehydes and ketones, and is active in various cancers such as pancreatic cancer [20, 21]. The second region with excessive African ancestral components is 2q37, in which *PDCD1* is involved in signaling of immune system and various diseases. The region showing the highest excessive European ancestral components is located on 3q13 which only harbors *LSAMP*, a candidate of tumor suppressor gene in human osteosarcomas and is associated with coronary artery disease [22, 23]. Interestingly, *EIF2AK2*, involved in influenza infection pathways [24, 25], is located ~200 Kb away from the peak (2p22) showing the second highest excessive European ancestral components, which suggested the resistance of influenza between African and European might be different. *PACRG*, located on 6q26, is associated with Parkinson's disease. However, the region 16q21 showing excessive European ancestral components does not contain any genes or known function elements.

The highest ancestry deviation among the six regions showing excess of African or European ancestral components is <0.026, which is much lower than that in any other study on New World admixed populations [10–12, 16–18]. For exampls, the African ancestral components' deviation exceeds 0.14 in Puerto Rican according to Tang et al. [11], which is more than fivefoldd higher than that in this study. We attribute the low ancestry deviation to the much larger sample size, denser markers, more confident phased data using trios, and more powerful statistical methods. Based on the maximum deviation regions, we estimated that the highest selection coefficient is <0.002 based on 14-generation CGF model. In fact, the selection coefficient in reality could be much lower than 0.002 considering the statistics and evolutionary noises. These results are reasonable considering the fact that there is much lower mortality of African Americans in recent 200 years compared with that during Atlantic slave trade.

4.3 African/European Ancestral Components in African Americans

The African and European ancestral chromosome segments in African Americans, inferred by HAPMIX, were collectively referred to as African and European ancestral components, respectively. The African and European ancestral components could be considered as an artificial African population (AAF) and an artificial European population (AEU), respectively, both residing in United States but with different evolution history. The genotype of AAF and AEU is credible because the accuracy of HAPMIX for ancestry inference is about 98 % based on simulated African Americans. Population differentiation (F_{ST}) between AEU and each putative European parental population was calculated, with CEU showing the lowest F_{ST} with AEU ($F_{ST\ [AEU\text{-}CEU]} = 0.0005$) among all putative parental populations. When 2,648 Caucasians from Illumina iControl database (referred to as Caucasian) were considered, $F_{ST\ [AEU\text{-}Caucasian]}$ was 0.0006, which was the second lowest. Among all indigenous African populations, YRI showed the lowest F_{ST} with AAF ($F_{ST\ [AAF\text{-}YRI]} = 0.0007$). The results consistent with our aforementioned analysis that YRI are CEU are populations that are most similar to AAF and AEU, respectively.

We performed an extensive forward-time simulation to explore the potential impact of genetic drift on the locus-specific population differentiations between AAF and YRI, as well as that between AEU and CEU. In this simulation, recombination was introduced according to the genetic map adapted from HapMap (release #22), and mutation was ignored given the short history of African Americans. The effective population sizes (N_e) of each population were obtained from the HapMap website. In particular, N_e for African, European, and African Americans were set to 17,094, 11,418, and 17,094, respectively. The African Americans were simulated based on the 14-generation CGF model in which gene flow from European parental population to the African Americans per generation was $\alpha = 0.017$. Finally, genotypes of 113 Europeans, 113 Africans, and 1,890 African Americans, as well as their SNP-specific ancestral status, were output to match the sample sizes of the data. Based on the primary simulation, we performed extended simulations by setting a bottleneck event in the first generation of African Americans, during which N_e for African Americans were reduced to 8,000, 5,000, 3,000, 2,000 and 1,000 respectively.

When the observed F_{ST} were compared with their counterpart between simulated populations under neutrality, different patterns showed up between European ancestry and African ancestry. For European ancestry, the observed F_{ST} between AEU and CEU (0.0005) was lower than that simulated (simulated $F_{ST\ [AEU\text{-}CEU]} = 0.0006$), and the genome-wide distribution of observed locus-specific F_{ST} did not deviate much from those simulated ($P = 0.042$, Kolmogorov–Smirnov test). However, for African ancestry, the observed F_{ST} between AAF and YRI (0.0007) was higher than the simulated value (simulated $F_{ST\ [AAF\text{-}YRI]} = 0.0006$), and genome-wide distribution of locus-specific $F_{ST\ [AAF\text{-}YRI]}$ was significantly

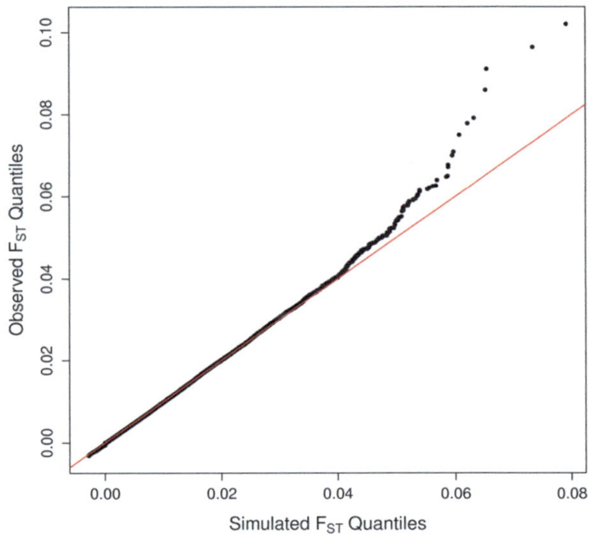

Fig. 4.4 Q–Q plot of locus-specific F_{ST} between AAF and YRI. This Q–Q plot compares empirical values on the vertical axis to simulated values on horizontal axis. *Red line* shows null distribution that F_{ST} of empirical data are the same as that of simulated

different from those simulated ($P < 2.2 \times 10^{-16}$, Kolmogorov–Smirnov test). In particular, Q–Q plot between the observed and simulated locus-specific $F_{ST\,[AAF\text{-}YRI]}$ showed an enrichment of SNPs with high F_{ST} at the tail of the observed locus-specific F_{ST} (Fig. 4.4), suggesting a possible role of natural selection. The results were essentially unchanged when bottlenecks for African immigrants were assumed in our simulations.

4.4 Identifying Highly Differentiated Regions Between AAF and Indigenous African

African Americans have experienced a completely different history compared with the indigenous population in Africa since the African immigrants left for America. The high mortality of African Americans during the slavery era [14, 15, 26] suggested a possible presence of natural selection. According to modern population genetics, demographic history influences the population difference (F_{ST}) of each locus similarly, if not equally [2, 27]. By contrast, the force of positive selection acts in a locus-specific manner and tends to increase F_{ST} [28], therefore, F_{ST} has been widely used to scan for signals of recent positive selection [29, 30]. In this study, we calculated the locus-specific F_{ST} between AAF and indigenous African population YRI at genome-wide. However, loci with extremely low minor allele frequency (MAF) may be subjected to sampling and statistical error, the SNPs with MAF < 0.05 in either AAF or YRI were removed from further analyses.

Genome-wide distribution of F_{ST} between AAF and YRI for 401,599 autosomal SNPs was presented similar to Manhattan plot (Fig. 4.5a). Although the

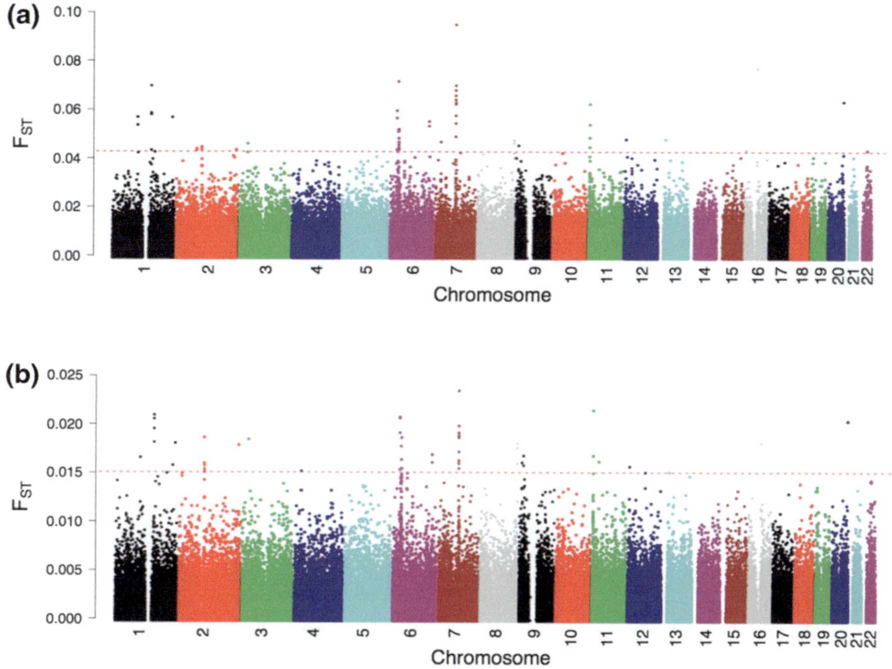

Fig. 4.5 Genomic distribution of F_{ST} between AAF and YRI (**a**) and genomic distribution of F_{ST} between rAfA and African Americans (**b**). The *dashed red horizontal line* indicates the cut-off threshold (99.99th percentile). Locus–specific F_{ST} between YRI and CEU were calculated when MAF > 0.05 in both populations. The rAfA was constituted according to the ancestry proportion of CEU and YRI under neutrality. Reprint from Jin et al. [55]

average population differentiation between AAF and YRI is low, a substantial fraction of SNPs were located at the right tail of the distribution (Fig. 4.4). The most differentiated SNPs (99.99th percentile; $F_{ST} > 0.0452$) are located in 14 regions (Table 4.2). Although it is reported that F_{ST} of individual-marker are too variable [28], each of the four most impressive regions (7q21, s6p21-22, 1q22, and 11p15) involves multiple highly differentiated SNPs and should be indicative. The most pronounced region showing population difference is 7q21, which contains two genes: *CD36* and *SEMA3C*. *CD36*, also known as *FAT*, is an integral membrane protein on cell surface and is involved in the transport of fatty acid, as well as fatty acid metabolism. *CD36* also mediates cytoadherence of *Plasmodium falciparum* parasitized erythrocytes, and it was reported that *CD36* has been subjected to strong positive selection for malaria or unknown pathogens [31–35]. *SEMA3C*, induced by *ADAMTS1*, promotes the migration of cancer cells [36]. The 6p21-22 region is much wider and contains human major histocompatibility complex (MHC). This region also showed an over-representation of African ancestral components in this study (although less than 3SD) and was reported under selection in various studies [11, 37]. *MUC1*, located on 1q22, is involved in signaling by PDGF and serves a protective function by binding to pathogens [38, 39]. *HBB* and

Table 4.2 Regions with highly differentiated allele frequency between AAF and YRI ($F_{ST} > 0.0452$)

Regions or SNPs	Position	Size (bp)	SNPs	Highest F_{ST}	Genes	Pathways	Related disease
1p21	chr1:100125058..100183875	58,817	2	0.0562	*AGL**	Metabolism of carbohydrate	Glycogen storage disease
1q22	chr1:153401959..153464086	62,127	4	0.0692	*THBS3**, *MUC1**, *MTX1*, *TRIM46*, *KRTCAP2*	Signaling by PDGF	Stomach cancer, breast cancer, osteosarcoma
rs12094201	chr1:236509336	1	1	0.0561	(*ZP4** 389 kb)	NA	Hypertension, non-alcoholic fatty liver
rs7642575	chr3: 31400165	1	1	0.0453	(*STT3B*, *OSBPL10** 149 kb)	NA	Peripheral arterial disease
6p21–p22	chr6:26554684..33961049	7,406,365	11	0.0711	*HLA-B**, *HLA-C*, *EHMT2**, *HLA-DPA1**, *HLA-DRB5*, *EHM*, *BTN3A3*, etc.	Signaling by GPCG, signaling in immune system, HIV infection, Diabetes pathway	HIV, Crohn's disease, rheumatoid arthritis, juvenile idiopathic arthritis, colorectal cancer, systemic sclerosis
6q25	chr6:151555551..151569258	13,707	2	0.0545	(*AKAP12** 40 kb)	Cell growth	Hypertension, hemorrhagic stroke
rs10499542	chr7: 22235870		1	0.04606	*RAPGEF5**	GTP/GDP-regulation	Thyroid stimulating hormone
7q21	chr7:79768487..80482597	714,110	10	0.0946	*CD36**, *SEMA3C*	Metabolism of lipids and lipoprotein	Metabolic syndrome, malaria
8q24	chr8:143754039..143758933	4894	2	0.04679	*PSCA**	NA	Prostate cancer, bladder cancer, gastric cancer

(continued)

Table 4.2 (continued)

Regions or SNPs	Position	Size (bp)	SNPs	Highest F_{ST}	Genes	Pathways	Related disease
11p15	chr11:5034229..5421456	387,227	3	0.0617	*HBB**, *HBD**, *HBE1**, *HBG2*, *OR51I1*, etc.	Signaling by GPCR	Sickle-cell disease, beta-thalassemia, malaria
rs4883422	chr12:7189594	1	1	0.04721	*CLSTN3*	NA	NA
rs6491096	chr13:25488362	1	1	0.04716	*ATP8A2*	NA	NA
rs1075875	chr16: 47595721	1	1	0.0766	(*CBLN1* 277 kb)	NA	NA
rs6015945	chr20:59319574	1	1	0.0627	*CDH4**	Cell junction organization	Alzheimer's disease

NA: not available. * Denotes the genes associated with diseases. Genes in parentheses are strong candidates out of the chromosome location but closest

HBD, located on 11p15, have been subjected to balancing selections because their mutations protected against malaria according to numerous studies [40, 41].

African Americans have higher risk for most major cancers and obesity-related diseases such as diabetes, hypertension, and prostate cancer compared with European Americans [42, 43]. Interestingly, many genes located in selection-candidate regions identified in this way are associated with African Americans' ethnic specific diseases such as hypertension, prostate cancer, and systemic sclerosis. Especially, one of the most significantly differentiated SNPs (rs2294008; $F_{ST} = 0.04561$) between AAF and YRI, located on 8q24, is a missense mutation c.57T > C (p.Met1Thr) in *PSCA* and was reported to be associated with gastric and bladder cancer [44, 45]. Many studies reported that multiple loci on 8q24 were associated with prostate cancer in African Americans [46–48]. We proposed a hypothesis that most of the genes associated with African Americans ethnic diseases may have played an important role in African Americans adapting to local environment and thus showing higher population differentiation between AAF and indigenous African population YRI. We think further analysis of the 8q24 region would provide new insights into the etiology and history of these cancers in African Americans.

Ingenuity pathway analysis (IPA) was particularly helpful in exploring the function and pathways of the selection-candidate genes in the context of higher-order cellular and molecular mechanisms. The 402 SNPs with the highest F_{ST} (99.90th percentile; $F_{ST} > 0.0287$) were subjected to IPA analysis, whose results showed that genes involved in metabolic diseases were the most significantly enriched ($P = 1.51 \times 10^{-16}$) among all function classes, followed by the genes involved in endocrine system disorder ($P = 2.23 \times 10^{-16}$), immunological diseases ($P = 9.30 \times 10^{-12}$), and genetic disorder ($P = 5.67 \times 10^{-11}$). Antigen presentation pathway was the most significantly enriched pathway ($P = 1.95 \times 10^{-4}$), followed by allograft rejection signaling ($P = 4.69 \times 10^{-3}$), graft-versus-host disease signaling ($P = 4.69 \times 10^{-3}$) and autoimmune thyroid diseases signaling ($P = 5.35 \times 10^{-3}$). All of the four aforementioned pathways are related to immune system, which might reflect a great environmental differentiation between Sub-Saharan Africa and North America. We also conducted IPA on 4,011 SNPs showing the highest F_{ST} (99.00th percentile; $F_{ST} > 0.0162$), which yielded results similar to that with $F_{ST} > 0.0287$ (99.90th percentile). And two additional pathways emerged: IL-9 signaling pathway ($P = 8.01 \times 10^{-3}$) and EGF signaling pathway ($P = 6.38 \times 10^{-3}$).

4.5 Reconstituted African Americans and Its Difference with African Americans

We simply reconstituted an African American population (rAfA) based on the genotypes of YRI and CEU. We then compared the rAfA genome with that of African Americans to detect the signatures of positive selection. The rationale behind is that the former evolve neutrally, while the latter have been subjected to possible positive selection. In particular, the allele frequencies of rAfA were estimated at each locus based on 78.35 % YRI and 21.65 % CEU assuming no natural selection. The signatures of natural selection identified in this way avoid potential errors introduced in ancestry inference. The genome-wide distribution of F_{ST} between rAfA and African Americans looks similar to that between AAF and YRI (Fig. 4.5b). Overall 81 % of the SNPs showing highest differential between AAF and YRI could be validated in the high differential SNPs between rAfA and African Americans. In particular, these major F_{ST} peaks between AAF and YRI are essentially the same as those between rAfA and African Americans, which indicated that almost all the signatures of selection identified by comparing rAfA and African Americans originated from AAF or YRI. We also used Caucasian instead of CEU to construct rAfA, with the genomic distribution of F_{ST} similar to that using CEU.

Next, we reconstructed a rAfA population using a set of putative parental populations. First, we constructed an African parental population of African Americans (APP) using 64 % Yoruba, 19 % Mandenka, and 14 % Bantu according to previously reported ancestral contribution to AAF [13]. Then we used genotypes of the APP and Caucasian to reconstruct rAfA. We obtained the genome-wide distribution of F_{ST} between rAfA and African Americans (Fig. 4.6). Although the population differentiation between this rAfA and African Americans is higher than that based on two aforementioned rAfAs, the genome-wide distribution of F_{ST} is similar to that using two pure parental populations.

4.6 Further Evidences for Positive Selection in African Ancestral Components

Since positive selection acts in a locus-specific manner and tends to increase F_{ST} [49], we hypothesized that positive selection preferentially acted upon functionally important loci over the others in the genome, thus leading to an enrichment of important functional SNPs in the high F_{ST} bin [9]. Here, we investigated the

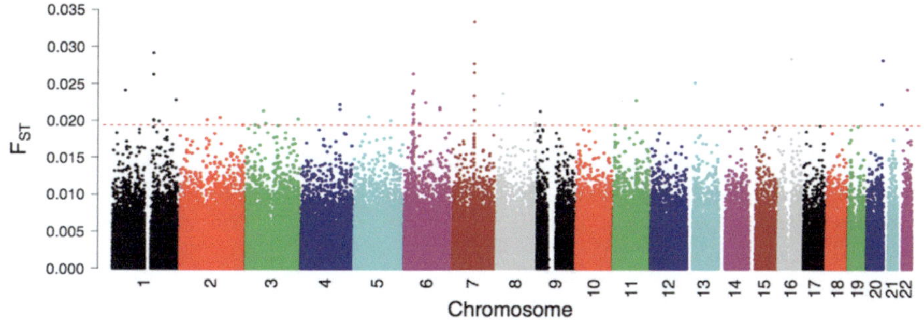

Fig. 4.6 Genome–wide distribution of F_{ST} between African Americans and rAfA. rAfA is constructed by Caucasian and APP (African parental population of African Americans constructed by YRI, Mandenka, and Bantu) according to estimated admixture proportion. The *dashed red horizontal line* indicates the cutoff threshold (99.99th percentile)

enrichment of different SNP categories among the high F_{ST} bin (top 1st percentile among all SNPs, $F_{ST} > 0.0164$) between AAF and YRI.

The SNPs were classified into non-genic, genic, intronic, 3'UTR, 5'UTR, synonymous, non-synonymous, coding, transcriptonic, near-gene-3, and near-gene-5 based on their location and function relative to the genes. The F_{ST} of functional SNP categories were not significantly different from that of non-genic SNPs. However, the proportion of genic SNPs among high F_{ST} bin (top 1st percentile among all SNPs, $F_{ST} > 0.0164$) is significantly higher than that of non-genic SNPs (χ^2 test, $P = 0.046$; Fig. 4.7). Notably, this excess is particularly marked for transcriptonic SNPs (χ^2 test, $P = 0.004$; Fig. 4.7). The proportion of synonymous SNPs in high F_{ST} bin was 1.22-fold higher than the expectation under neutrality, which could attribute to the linkage disequilibrium of these SNPs with the selected loci (a phenomenon known as hitchhiking). Similar observations could be made when the thresholds for high F_{ST} bin were set at top 5 % ($F_{ST} > 0.0083$) or top 0.1 % ($F_{ST} > 0.0304$), and the conclusions were essentially changed when the SNPs with MAF >0.05 in both YRI and AAF were examined. The significant enrichment of high F_{ST} loci in the SNP categories with genetic functions supports the presence of positive selection either in AAF or in YRI, or both.

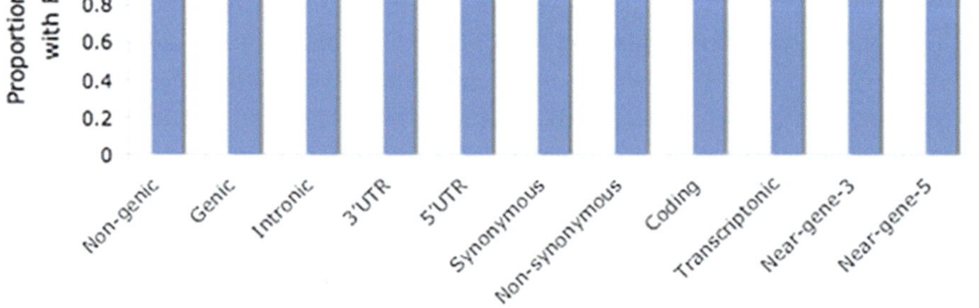

Fig. 4.7 Enrichment of high F_{ST} loci for different SNP categories. Observed excess of high F_{ST} loci in different SNP categories, with respect to non-genic class, among high F_{ST} bin (99th percentile; $F_{ST} > 0.0164$). The values on the bar are p-values of χ^2 tests. "NS" stands for "not significant". Reprint from Jin et al. [55]

4.7 Relaxed Natural Selection in African Americans

Among the selection-candidate genes detected by genome-wide locus-specific F_{ST} between AAF and indigenous African population YRI, five genes (*CD36*, *HBB*, *HBD*, *HLA-B*, *HLA-DR*) have been reported subjected to natural selection due to malaria [50, 51]. Because of the strong selection pressure of malaria, loss-of-function or abnormality of these genes was supposed to increase the survival rate of individual living in Africa. Some mutations in these genes have reached much higher frequencies in Africans compared with that in areas of low incidence of malaria. However, these mutations that defend malaria could become disadvantage in African Americans because the malaria was no longer a strong selection force in North America and these mutations even lead to morbidity or mortality [52]. We hypothesize that frequencies of these mutations would have decreased in AAF compared with those in indigenous Africans due to their disadvantage in African Americans.

We examined this hypothesis using our empirical data. We examined the SNPs that are in high-linkage disequilibrium with these functional mutations instead of these mutations themselves because the functional mutations were not genotyped in the data we used. It is well known that rs3211938 is a nonsense mutation c.1389T > G (p.Tyr325X) in *CD36* and has been subjected to natural selection because of malaria or some unknown environmental factors in Africans [32–35]. We found three SNPs that are highly differentiated between AAF and YRI (Table 4.3) on 7q21 are strongly linked with rs3211938 (each with $r^2 > 0.4$ in YRI). Interestingly, we did observe the frequencies of the alleles linked with derived allele rs3211938(G) were much lower in AAF compared with that in YRI (Table 4.3). Furthermore, rs334 is a missense mutation c.70A > *T* (p.Glu7Val) in *HBB*, which leads to sickle-cell disease (MIM 603903) [40, 41, 53], one of the most well-studied genetic disorders. rs7952293, a SNP with high F_{ST} between AAF and YRI, was strongly linked with rs334 ($r^2 = 0.237$ in YRI). In particular, the haplotype constructed by rs7952293(A) and rs334(T) accounted for 86.67 % of the haplotypes containing rs334(T) in YRI. We observed that the frequency of rs7952293(A) in AAF (0.2261) was much lower than that in YRI (0.3172), which also supports the hypothesis that frequencies of alleles protecting against malaria in AAF are lower than those in indigenous Africans. The other three genes were not examined using the same procedure because the frequencies of mutations on these genes are too low to find strong-linked representative SNPs.

Table 4.3 SNPs showing strong linkage with rs3211938(G) in CD36

dbSNP_id	Alleles	Allele linked with rs3211938(G)	Frequency (YRI)	Frequency (AAF)	F_{ST} (AAF-YRI)	r^2 (YRI)
rs10216027	C/T	T	0.29	0.16	0.0569	0.514
rs1404315	A/C	C	0.32	0.16	0.0693	0.489
rs1722504	C/T	C	0.38	0.22	0.0652	0.403

4.8 Disease Susceptibility Genes Enriched Among Signatures of Selection

We identified a large number of genes with highly differentiated allele frequencies between AAF and YRI. IPA Analysis of SNPs with high population differentiation between AAF and YRI showed that genes involved in metabolic diseases ($P = 1.51 \times 10^{-16}$), endocrine system disorder ($P = 2.23 \times 10^{-16}$), immunological diseases ($P = 9.30 \times 10^{-12}$), and genetic disorder ($P = 5.67 \times 10^{-11}$) were significantly enriched. African Americans have higher risk for most major cancers and obesity-related diseases, such as diabetes, hypertension, and prostate cancer, compared with European Americans [42, 43, 54]. Interestingly, many genes located in selection-candidate regions identified in this study, such as *PSCA*, *ZP4*, *AKAP12,* are associated with African Americans' ethnic specific diseases such as hypertension, prostate cancer, and systemic sclerosis. Five genes (*CD36*, *HBB*, *HBD*, *HLA-B*, *HLA-DR*), whose mutations lead to protect against malaria, are also located in the highly differentiated regions between AAF and YRI. Especially, one of the most significantly differentiated SNPs (rs2294008; $F_{\mathrm{ST}} = 0.04561$) between AAF and YRI, located on 8q24, is a missense mutation c.57T $>$ *C* (p.Met1Thr) in *PSCA* and was reported to be associated with gastric and bladder cancer [44, 45]. Many studies reported that multiple loci on 8q24 were associated with prostate cancer in African Americans [46–48]. We proposed a hypothesis that most of the genes associated with African Americans ethnic diseases may have played an important role in African Americans adapting to local environment and thus showing higher population differentiation between AAF and indigenous African population YRI.

4.9 Implication for the Future Study

Admixed populations such as African Americans provide a unique opportunity for studying very recent natural selection, as their genomes are donated by long-diverged continental ancestries and have been subjected to novel environmental challenges. The first approach, detecting excessive or decreased ancestry contribution from its parental populations, has been used in several studies [10–12]. However, natural selections that act on the parental population before population admixture could not influence the parental population's genetic contribution. Therefore, we developed a new strategy that would detect signatures of natural selection since the African ancestors of African Americans left for America, including nature selection both pre-and post-admixture. The selection-candidate genes identified by the new strategy were also confirmed by another evidence, which compared locus-specific population differentiation between rAfA and African Americans, although the two strategies are not completely independent.

This study takes advantage of both large sample size and high-density genome-wide data. The results showed that the maximum ancestry deviation, either excess of African or European ancestry, was <2.6 %, Therefore, detecting such weak signatures of selection in admixed population is a big challenge and much large sample size was needed for distinguishing the real signals from the ancestry deviation caused by genetic drift and sampling error. We propose that any study trying to detect signatures of selection in recent admixed populations should collect at least thousands of samples in the future. Our new strategy has detected a lot of genes associated with African Americans' specific high-risk diseases such as hypertension and prostate cancer, and has also detected five genes whose mutations protect against malaria. We suggest that our new approach is much more powerful in detecting natural selection both before and after the establishment of African Americans, and can be applied to other admixed populations such as Latinos.

References

1. Balaresque PL, Ballereau SJ, Jobling MA. Challenges in human genetic diversity: demographic history and adaptation. Hum Mol Genet. 2007;16(R2):R134–9.
2. Kimura M. The neutral theory of molecular evolution. Cambridge: Cambridge University Press; 2003.
3. Sabeti PC, et al. Genome-wide detection and characterization of positive selection in human populations. Nature. 2007;449:913–8.
4. Akey JM. Constructing genomic maps of positive selection in humans: where do we go from here? Genome Res. 2009;19:711–22.
5. Balaresque PL, Ballereau SJ, Jobling MA. Challenges in human genetic diversity: demographic history and adaptation. Hum Mol Genet. 2007;16(R2):R134–9.
6. Nielsen R, Hellmann I, Hubisz M, Bustamante C, Clark AG. Recent and ongoing selection in the human genome. Nat Rev Genet. 2007;8:857–68.
7. Hancock AM, et al. Adaptations to climate in candidate genes for common metabolic disorders. PLoS Genet. 2008;4:e32.
8. Pickrell JK, et al. Signals of recent positive selection in a worldwide sample of human populations. Genome Res. 2009;19:826–37.
9. Barreiro LB, Laval G, Quach H, Patin E, Quintana-Murci L. Natural selection has driven population differentiation in modern humans. Nat Genet. 2008;40:340–5.
10. Bryc K, et al. Genome-wide patterns of population structure and admixture in West Africans and African Americans. Proc Natl Acad Sci U S A. 2010;107:786–91.
11. Tang H, et al. Recent genetic selection in the ancestral admixture of Puerto Ricans. Am J Hum Genet. 2007;81:626–33.
12. Basu A, et al. Genome-wide distribution of ancestry in Mexican Americans. Hum Genet. 2008;124:207–14.
13. Zakharia F, et al. Characterizing the admixed African ancestry of African Americans. Genome Biol. 2009;10:R141.
14. Thomas, H. The slave trade: the story of the atlantic slave trade. Simon Schuster 1440–1870(1999).
15. Stannard, D. American Holocaust. Oxford University Press (1993).
16. Workman PL, Blumberg BS, Cooper AJ. Selection, gene migration and polymorphic stability in A U. S. white and negro population. Am J Hum Genet. 1963;15:429–37.
17. Blumberg BS, Hesser JE. Loci differentially affected by selection in two American black populations. Proc Natl Acad Sci USA. 1971;68:2554–8.

18. Reed TE. Caucasian genes in American Negroes. Science. 1969;165:762–8.
19. Price AL, et al. Long–range LD can confound genome scans in admixed populations. Am J Hum Genet. 2008;83:132–5.
20. Praml C, et al. Genetic variation of Aflatoxin B1 aldehyde reductase genes (AFAR) in human tumour cells. Cancer Lett. 2008;272:160–6.
21. Cui Y, et al. Proteomic analysis of pancreatic ductal adenocarcinoma compared with normal adjacent pancreatic tissue and pancreatic benign cystadenoma. Pancreatology. 2009;9:89–98.
22. Yen CC, et al. Identification of chromosomal aberrations associated with disease progression and a novel 3q13. 31 deletion involving LSAMP gene in osteosarcoma. Int Joncology. 2009;35:775.
23. Kresse, S.H. et al. LSAMP, a novel candidate tumor suppressor gene in human osteosarcomas, identified by array comparative genomic hybridization. Genes Chromosom Cancer 48(2009).
24. McAllister CS, et al. Mechanisms of protein kinase PKR–mediated amplification of beta interferon induction by C protein–deficient measles virus. J Virol. 2010;84:380–6.
25. Pereira RM, et al. Novel role for the double–stranded RNA–activated protein kinase PKR: modulation of macrophage infection by the protozoan parasite Leishmania. Faseb J. 2010;24:617–26.
26. Meltzer, M. Slavery: a world history. Da Capo Press (1993).
27. Weir BS, Cockerham CC. Estimating F–statistics for the analysis of population structure. Evolution. 1984;38:1358–70.
28. Weir BS, Cardon LR, Anderson AD, Nielsen DM, Hill WG. Measures of human population structure show heterogeneity among genomic regions. Genome Res. 2005;15:1468.
29. Akey JM, Zhang G, Zhang K, Jin L, Shriver MD. Interrogating a high–density SNP map for signatures of natural selection. Genome Res. 2002;12:1805.
30. Oleksyk TK, Smith MW, O'Brien SJ. Genome–wide scans for footprints of natural selection. Philos Trans R Soc Lond B Biol Sci. 2010;365:185–205.
31. Baruch DI, Gormely JA, Ma C, Howard RJ, Pasloske BL. Plasmodium falciparum erythrocyte membrane protein 1 is a parasitized erythrocyte receptor for adherence to CD36, thrombospondin, and intercellular adhesion molecule 1. Proc Natl Acad Sci U S A. 1996;93:3497–502.
32. Erdman, L.K. et al. CD36 and TLR Interactions in Inflammation and Phagocytosis: Implications for Malaria. J Immunol (2009).
33. Fry AE, et al. Positive selection of a CD36 nonsense variant in sub–Saharan Africa, but no association with severe malaria phenotypes. Hum Mol Genet. 2009;18:2683–92.
34. Aitman TJ, et al. Malaria susceptibility and CD36 mutation. Nature. 2000;405:1015–6.
35. Omi K, et al. CD36 polymorphism is associated with protection from cerebral malaria. Am J Hum Genet. 2003;72:364–74.
36. Esselens C, et al. The cleavage of semaphorin 3C induced by ADAMTS1 promotes cell migration. J Biol Chem. 2010;285:2463–73.
37. Garrigan D, Hedrick PW. Perspective: detecting adaptive molecular polymorphism: lessons from the MHC. Evolution. 2003;57:1707–22.
38. Davila S, et al. New genetic associations detected in a host response study to hepatitis B vaccine. Genes Immun. 2010;11:232–8.
39. Li Y, Dinwiddie DL, Harrod KS, Jiang Y, Kim KC. Anti–inflammatory effect of MUC1 during respiratory syncytial virus infection of lung epithelial cells in vitro. Am J Physiol Lung Cell Mol Physiol. 2010;298:L558–63.
40. Ashley–Koch A, Yang Q, Olney RS. Sickle hemoglobin (HbS) allele and sickle cell disease: a HuGE review. Am J Epidemiol. 2000;151:839–45.
41. Wood ET, Stover DA, Slatkin M, Nachman MW, Hammer MF. The beta –globin recombinational hotspot reduces the effects of strong selection around HbC, a recently arisen mutation providing resistance to malaria. Am J Hum Genet. 2005;77:637–42.
42. Jemal A, et al. Cancer statistics. CA Cancer J Clin. 2006;56:106–30.

43. Goran MI. Ethnic–specific pathways to obesity–related disease: the Hispanic vs. Afr Am paradox. Obesity (Silver Spring). 2008;16:2561–5.
44. Sakamoto H, et al. Genetic variation in PSCA is associated with susceptibility to diffuse–type gastric cancer. Nat Genet. 2008;40:730–40.
45. Wu X, et al. Genetic variation in the prostate stem cell antigen gene PSCA confers susceptibility to urinary bladder cancer. Nat Genet. 2009;41:991–5.
46. Freedman ML, et al. Admixture mapping identifies 8q24 as a prostate cancer risk locus in African–American men. Proc Natl Acad Sci U S A. 2006;103:14068–73.
47. Al Olama AA, et al. Multiple loci on 8q24 associated with prostate cancer susceptibility. Nat Genet. 2009;41:1058–60.
48. Yeager M, et al. Identification of a new prostate cancer susceptibility locus on chromosome 8q24. Nat Genet. 2009;41:1055–7.
49. Nielsen R. Molecular signatures of natural selection. Annu Rev Genet. 2005;39:197–218.
50. Kwiatkowski D. The molecular genetic approach to malarial pathogenesis and immunity. Parassitologia. 1999;41:233–40.
51. Kwiatkowski DP. How malaria has affected the human genome and what human genetics can teach us about malaria. Am J Hum Genet. 2005;77:171–92.
52. Platt OS, et al. Mortality in sickle cell disease. Life expectancy and risk factors for early death. N Engl J Med. 1994;330:1639–44.
53. Winichagoon P, Fucharoen S, Chen P, Wasi P. Genetic factors affecting clinical severity in beta–thalassemia syndromes. J Pediatr Hematol Oncol. 2000;22:573–80.
54. Smith MW, O'Brien SJ. Mapping by admixture linkage disequilibrium: advances, limitations and guidelines. Nat Rev Genet. 2005;6:623–32.
55. Jin W, et al. Genome–wide detection of natural selection in African Americans pre– and post–admixture. Genome Res. 2012;22:519–27.

Chapter 5
Complex Selective Forces Shaping the Genes Underlying Human Diseases

Abstract Traditionally, genetic disorders have been classified as either monogenic or complex diseases. This nosology has greatly benefited genetic counseling and the development of gene mapping strategies. However, based on two well-established databases, we showed that 54 % (524 of 968) of the monogenic diseases genes were also involved in complex diseases. The genes underlying both monogenic and complex diseases were referred to as MC genes. Analyses of human polymorphic data, specie divergence data, copy number variation (CNV) data, gene expression data, and interaction network showed that MC genes are subject to both strong positive and strong purifying selections, simultaneously; while genes only underlying monogenic diseases and complex diseases were only subjected to strong purifying selection and strong positive selection, respectively.

Keywords Monogenic disease · Complex disease · Disease gene · Natural selection · Evolution · Copy number variation (CNV)

5.1 Genes Underlying Genetic Diseases

Genetics is believed to play an important role in all kinds of human diseases, either determining the occurrence of a disease or influencing the susceptibility and the severity of a disease. In some cases, a single genetic mutation is sufficient to lead to a disease, while in the other cases a disease is influenced by multiple genetic variants and environmental factors. Therefore, genetic disorders are typically classified into monogenic diseases and complex diseases, which has greatly benefited the development of gene mapping strategies and genetic counseling [1–4]. Monogenic diseases, such as sickle cell disease and cystic fibrosis, are caused by single-gene defects and their transmissions follow Mendel's law.

W. Jin, *Admixture Dynamics, Natural Selection and Diseases in Admixed Populations*, Springer Theses, DOI 10.1007/978-94-017-7408-6_5

Discovery of the genetic variations underlying monogenic diseases was pretty successful because of the implement of pedigree-based linkage analysis. The genes underlying most monogenic diseases were identified even before the completion of the Human Genome Project (HGP) [1, 5–7]. Studies on monogenic diseases have greatly contributed to our understanding of gene function and gene regulation mechanisms in these diseases, as well as facilitated the development of effective diagnostic and therapeutic approaches [5]. Although the monogenic disease altogether only affected only a small fraction of the general population, the argument for studying monogenic diseases was that they might reveal functional pathways and networks that might be involved in complex diseases [1, 5].

However, studies on monogenic diseases provided little insights into the genetic and biological basis of complex diseases [1, 4]. On the other hand, complex diseases, including most common diseases such as diabetes and cancers, are associated with the effects of multiple genes and environmental factors and are prevalent in the general population. During the past 6 years, advances in genomic-related technology have lured many scientists to tackle complex diseases. Genome-wide association studies (GWAS) assaying high-density SNPs in thousands of individuals have reproducibly identified thousands of genes associated with complex diseases [8, 9]. These studies not only identified new loci associated with complex diseases, but also provided many novel functional and biological insights. However, among those heritable components of complex disease, only a small fraction has been explained, and it remains a big challenge to elucidate underlying mechanisms between the associated variants and phenotypic traits [10].

Elucidating the evolution of disease genes is expected to shed light on the origin of genetic diseases, which is still one of the greatest challenges for medical genetics [11, 12]. Comparative analyses of genes underlying disease and genes not associated with any disease have been extensively performed [12–18]. In addition, several studies compared the characteristics of complex disease genes and monogenic disease genes from some aspects [19–22]. Based on these studies, the variants underlying monogenic disease have a very strong tendency to occur at highly conserved amino acid positions in proteins, suggesting that they generally have a severe impact on the function of the protein. It is consensus that genes underlying monogenic disease are subjected to strong purifying selection and are more conserved than those underlying complex diseases [19, 20]. By separating the non-disease genes into housekeeping and other-no-disease genes, it was found that housekeeping genes are the most conserved among all categories of genes, whereas other-no-disease genes show an evolutionary rate intermediate between genes underlying monogenic diseases and complex diseases. Although genes underlying complex disease have a higher number of interacting partners than that of monogenic disease genes and other genes, the reasons for their higher evolutionary rate require explanation [21].

5.2 Monogenic Disease Genes Are More Likely to Be Involved in Complex Diseases

Although most of the genetic variations underlying complex disease and monogenic disease may be different from each other, some genes involved in both of them have been reported [23–25]. It is not clear how many genes underlie both monogenic and complex diseases. We first investigated the occurrence of complex disease genes among the total 968 monogenic disease genes in hand-curated Mendelian Inheritance in Man database (hOMIM) [20], a database of high-penetration diseases caused by single mutation. Among the 3,117 complex disease genes from genetic association database (GAD) [26, 27], monogenic disease genes were significantly over-represented (524 of 968) compared with non-monogenic disease genes by 8.2-fold (Fig. 5.1; $P < 2.2 \times 10^{-16}$, χ^2 test), which also indicated that complex disease genes were significantly over-represented in monogenic disease genes. The strong co-occurrence of monogenic disease genes and complex disease genes suggested that monogenic disease genes are more likely to be involved in complex diseases and vice versa. The 524 genes implicated in both complex diseases and monogenic diseases are referred to as MC genes in the subsequent analyses. In addition, the 444 genes underlying monogenic diseases but not complex disease are simply referred to as monogenic genes, while the 2,593 genes underlying complex diseases but not monogenic diseases are simply referred to as complex genes.

The strong co-occurrence of genes underlying monogenic and complex diseases could be made when other datasets of disease genes were investigated. In particular, overall 1,799 genes implicated in complex diseases were obtained from GWAS catalog [8], a database including only those GWAS with at least 100,000 SNPs in the initial stage. We found 20.1 % (195 of 968) of the monogenic genes were in GWAS catalog, which was 5.1 times higher than the proportion of non-monogenic genes in GWAS catalog ($P < 2.2 \times 10^{-16}$, Fisher's exact test).

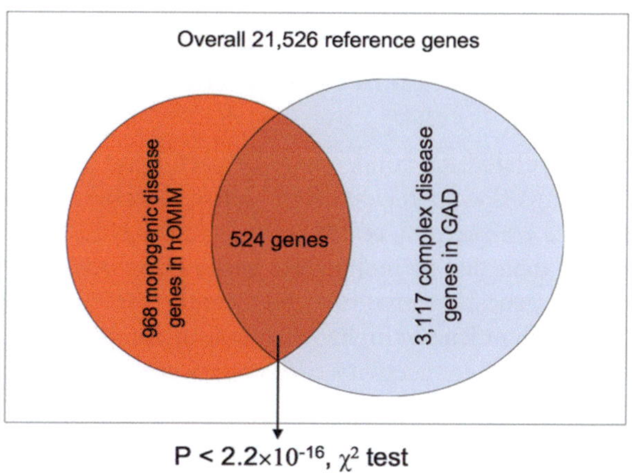

Fig. 5.1 Monogenic disease genes are more likely to be involved in complex diseases. Overall 54.1 % of (524 of 968) the monogenic disease genes are also genetic risk factors for complex diseases. χ^2 test was performed to examine whether monogenic disease genes were significantly enriched in complex diseases genes. Adapted from Jin et al. [60]

In another case, overall 3,261 monogenic disease genes were confirmed by at least two studies in OMIM [6, 7] and were used to represent monogenic disease genes. We found 41.3 % (1,288 of 3,117) of the complex disease genes were in OMIM, which was 3.8-fold higher than the proportion of non-complex disease genes in OMIM ($P < 2.2 \times 10^{-16}$, Fisher's exact test).

5.3 MC Genes Are Associated with More Diseases and Phenotypes

It is obvious that each MC gene is associated with at least two different diseases (one monogenic disease and one complex disease), we are interested in the exact number of diseases associated with each MC gene in hOMIM and GAD, respectively. To achieve this, we first classified the 968 monogenic disease genes into 524 MC genes and 444 monogenic genes (monogenic but not complex disease genes). We found each MC gene was associated with 1.77 ± 0.059 (mean ± SEM) monogenic diseases on average, significantly higher than that of a monogenic gene (1.35 ± 0.043, mean ± SEM; $P = 2.8 \times 10^{-6}$, Wilcoxon test). We further classified the 3,117 complex disease genes into 524 MC genes and 2,593 complex genes (complex but not monogenic disease genes). Interestingly, each MC gene was associated with 5.83 ± 0.50 (mean ± SEM) complex diseases on average, significantly higher than that of a complex gene (2.89 ± 0.14, mean ± SEM; $P < 2.2 \times 10^{-16}$, Wilcoxon test). In a nutshell, each MC gene was associated with 5.83 complex diseases and 1.77 monogenic diseases on average, much higher than that of a complex gene or a monogenic gene, respectively. Similar conclusions could be drawn when each gene associated phenotype was investigated (these phenotypes were defined in hOMIM and GAD, respectively).

5.4 MC Genes Involved in More Complex Interaction Network Than that of Monogenic or Complex Genes

We calculated the connectivity of each gene based on the protein–protein interaction network constructed using human protein reference database (HPRD) [28]. The average connectivity of MC genes (12.00 ± 1.02; mean ± SEM) was significantly higher than that of monogenic genes ($P = 3.43 \times 10^{-7}$, Wilcoxon test) and non-disease genes (referred to as ND genes; $P < 2.2 \times 10^{-16}$, Wilcoxon test) (Fig. 5.2), but not significantly higher than that of complex genes ($P = 0.058$, Wilcoxon test). The average connectivity of complex genes (10.96) was significantly higher than

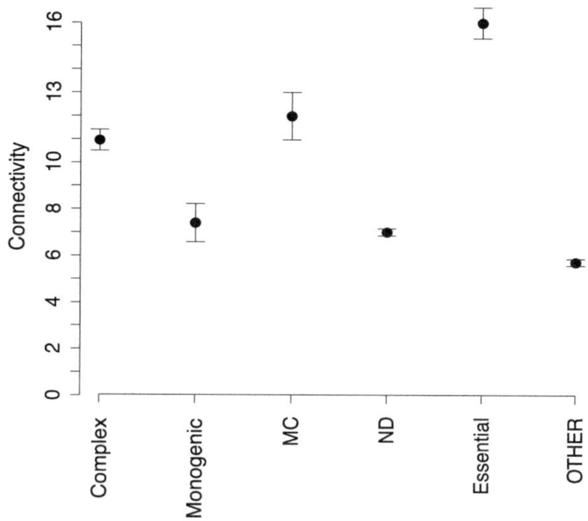

Fig. 5.2 Comparison of connectivity among different gene categories. The connectivity of each gene was calculated based on HPRD network, the protein–protein interaction network constructed from human protein reference database (HPRD). *Error bars* represent the standard error of the mean. Adapted from Jin et al. [60]

that of monogenic genes ($P = 9.57 \times 10^{-6}$, Wilcoxon test), consistent with the reports that connectivity of complex genes was significantly higher than that of monogenic genes [21, 29].

We used a set of 1,520 genes for which knockouts were either inviable or caused sterility in mouse to represent human essential genes [30]. By further separating ND genes into essential genes and OTHER genes, we found that the connectivity of an essential gene was the highest (16.00 ± 0.67; mean ± SEM) while that of an OTHER gene was the lowest (5.74 ± 0.14; mean ± SEM), with the connectivity of a MC gene (12.00), a complex gene (10.96), and a monogenic gene (7.40) ranging in between (Fig. 5.2). The median connectivity of each gene category ranks in the same order, i.e., essential genes (8), MC genes (6), complex genes (5), monogenic genes (4), and OTHER genes (3). These observations are consistent with those of some studies [29, 31], but not with that in a recent study by Podder and Ghosh [21] which, using only 83 complex (polygenic) disease genes, reported that complex disease genes had higher connectivity than housekeeping genes.

It is known that essential genes tend to encode highly interconnected proteins and resemble 'hubs' in yeast [32]. Whether disease genes tend to encode hub genes in human has been discussed in various studies [31]. Our analysis showed that the proportion of hub genes in disease genes was higher than that in OTHER genes, but lower than that in essential genes. Meanwhile, MC genes and complex genes contain higher proportion of hub genes than monogenic genes. The results based on network constructed using yeast two-hybrid database (YTH) [31], or combined dataset of HPRD and YTH (HPRD-YTH), were essentially the same as the aforementioned observations in HPRD.

5.5 MC Genes Have the Longest Coding Sequence (CDS)

Characteristics of gene structures were investigated to predict human disease genes in various studies [14, 18]. Here we examined coding sequence (CDS) length, gene length, 3′-UTR, 5′-UTR, and transcript count of the five gene categories. We found that the average CDS length of each disease gene category was significantly longer than that of OTHER genes ($P < 3.03 \times 10^{-12}$, Wilcoxon test) (Fig. 5.3a, b), which confirmed previous studies reporting that disease genes encoded longer proteins compared with non-disease genes [14, 18]. Furthermore, we observed that the average CDS length of MC genes was the longest (2,396 ± 237nt; mean ± SEM) in all gene categories, and even significantly longer than that of essential genes (P = 0.0084, Wilcoxon test). Then, we calculated the length of each gene based on its physical position in human reference genome (from gene start to gene end). We observed that average gene length of

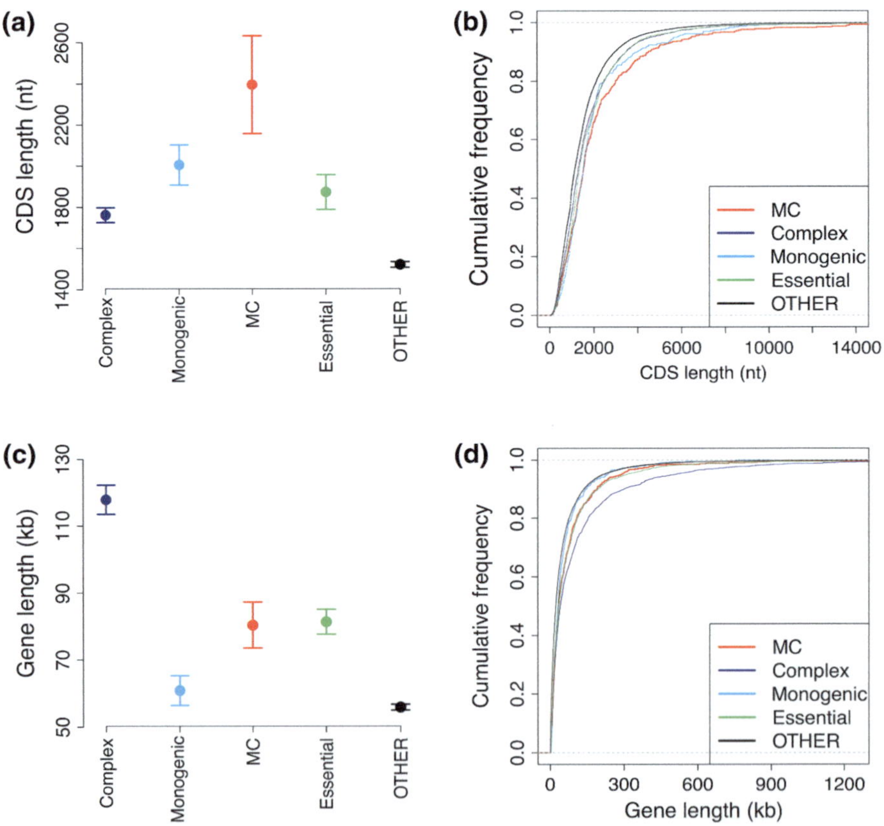

Fig. 5.3 Gene structures among different gene categories. *Error bars* represent the standard error of the mean. **a** Comparison of CDS length. **b** Cumulative distribution of CDS length. **c** Comparison of gene length. **d** Cumulative distribution of gene length. Adapted from Jin et al. [60]

each disease gene category was also significantly longer than that of OTHER genes ($P < 9.64 \times 10^{-7}$, Wilcoxon test) (Fig. 5.3c, d). However, instead of MC genes, the average gene length of complex genes was the longest among all gene categories, indicating that the compactness of MC genes was higher than that of complex genes. Gene length of MC genes (80,476nt ± 6,906nt; mean ± SEM) is not significantly different from that of essential genes ($P = 0.54$, Wilcoxon test), while significantly higher than those of monogenic genes and OTHER genes (Fig. 5.3c, d).

On the other hand, the average transcript count of each category of disease gene was significantly higher than that of OTHER genes ($P < 1.23 \times 10^{-10}$, Wilcoxon test), which may explain why many mutations altering the splicing sites are more likely to cause diseases [33]. Moreover, MC genes had the highest average transcript count (7.62 ± 0.30; mean ± SEM) among the five gene categories. The average 5′ UTR length of essential genes (229nt ± 6nt; mean ± SEM) was the longest among all gene categories ($P < 7.4 \times 10^{-5}$, Wilcoxon test), while that of MC genes (199nt ± 9nt; mean ± SEM) was not significantly different from other gene categories except essential genes. The average 3′ UTR length of MC genes (440nt ± 216nt; mean ± SEM) was the shortest among all categories. For example, MC genes showed shorter 3′ UTR than monogenic genes (1158nt ± 353nt; mean ± SEM) ($P = 0.045$, Wilcoxon test).

5.6 MC Genes Have the Highest Tissue Specificity

Gene expression variation is an important mechanism underlying susceptibility to complex diseases [34, 35]. Although gene expression patterns between healthy and pathological tissues have been extensively studied [35, 36], few studies have been conducted to systematically explore the gene expression patterns of disease genes in various healthy tissues. We analyzed the data of human U133A/GNF1H Gene Atlas including basal expression signal intensity of 44,776 probe sets across 84 human tissues. Our analyses showed that gene expression levels of the three disease gene categories and essential genes were significantly higher than that of OTHER genes on average ($P < 7.1 \times 10^{-9}$, Wilcoxon test), although the expression levels of the four gene categories were not significantly different from each other (Fig. 5.4a). S_{max} represents the highest expression signal of a gene across all tissues. Analysis showed that average S_{max} of MC genes (1568S ± 177S; mean ± SEM) was the highest among all gene categories (Fig. 5.4b), and was much higher than that of OTHER genes (almost fourfold).

Then we measured the tissue specificity of each gene using tissue specificity index τ [37], ranging from 0 to 1. Our analysis showed that average tissue specificity of each disease gene category was significantly higher than those of essential genes and OTHER genes ($P < 1.3 \times 10^{-5}$, Wilcoxon test; Fig. 5.4c, d), which supported previous studies reporting that highly differentially expressed genes are more likely to be associated with diseases [38, 39]. Especially, tissue specificity of MC genes ($0.57\tau \pm 0.0093\tau$; mean ± SEM) was much higher than that of any other

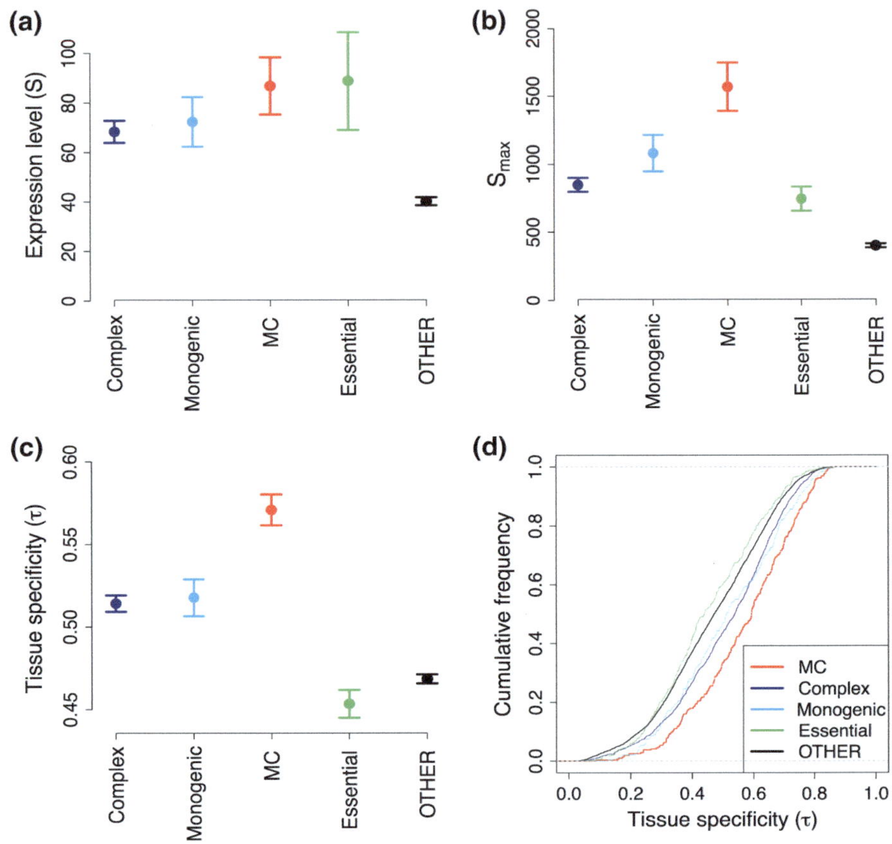

Fig. 5.4 Gene expression patterns showing by different statistics. *Error bars* represent the standard error of the mean. **a** Comparison of gene expression level. **b** Comparison of S_{max} (the highest expression signal of each gene across all tissues). **c** Comparison of tissue specificity. **d** Cumulative distribution of tissue specificity among different gene categories. Adapted from Jin et al. [60]

gene categories ($P < 3.5 \times 10^{-5}$, Wilcoxon test; Fig. 5.4c, d), although their expression level (86.85S ± 11.52S; mean ± SEM) was very similar to that of essential genes. Tissue specificities of complex genes and monogenic genes were similar to each other, and both were significantly lower than that of MC genes and significantly higher than those of essential genes and OTHER genes ($P < 1.3 \times 10^{-5}$, Wilcoxon test). Since both MC genes and complex genes had higher tissue specificity and transcript count than OTHER genes, it may suggest that complex disease genes might be involved in more complex gene regulation mechanism, which may explain the recent discovery that diseases associated SNPs identified by GWAS are more likely to be expression quantitative trait loci (eQTLs) [40].

Although housekeeping genes were thought to be similar to the essential genes, the approaches for identification of housekeeping genes and essential genes usually were completely different. In brief, essential genes are genes with knockout lethal, while housekeeping genes are genes with few expression variations in all

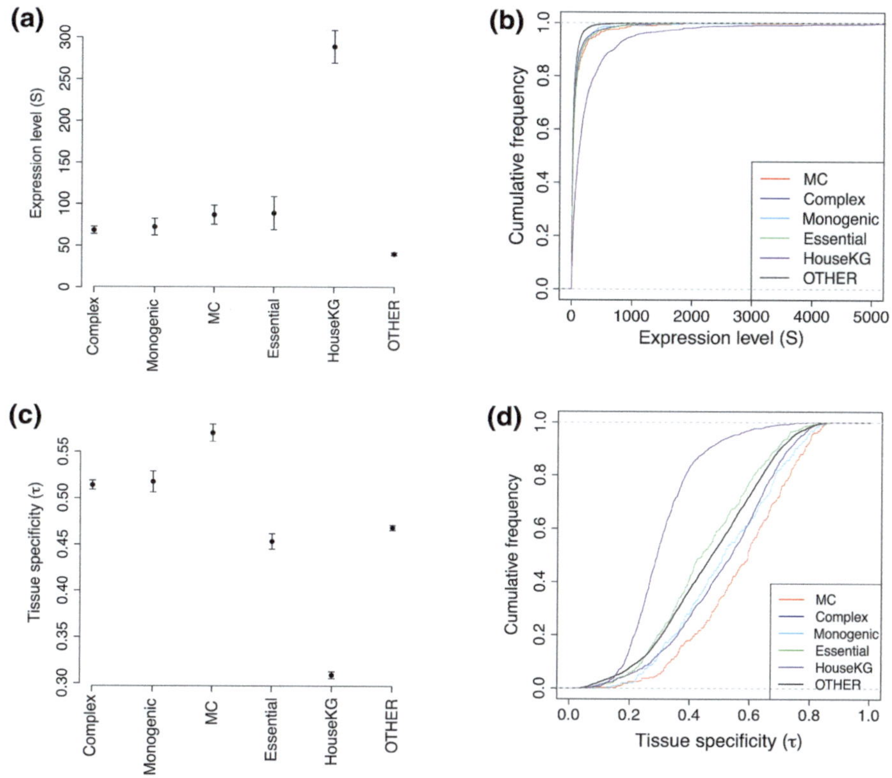

Fig. 5.5 Gene expression patterns including housekeeping genes. *Error bars* represent the standard error of the mean. **a** Comparing gene expression level. **b** Cumulative distribution of expression level in different gene categories. **c** Comparing tissue specificity. **d** Cumulative distribution of tissue specificity in different gene categories

the cell types. When 1,534 housekeeping genes from Tu et al. [12] were compared with all the other gene categories, we found that they had the highest average expression level and the lowest average tissue specificity among all gene categories (Fig. 5.5), which was consistent with previous reports [15, 41]. Although housekeeping genes have similar characteristics as essential genes and are thought as suitable substitutes for essential genes in most studies [12, 42], their expression levels and tissue specificities are, however, significantly different.

5.7 Recent Natural Selection on Different Gene Categories

With most mutations being lethal or seriously deleterious, essential genes and genes underlying monogenic disease have been reported under strong purifying selection [14, 43, 44]. Although some studies showed that genes underlying complex disease have been subjected to positive selection [20], there is no consensus on the genetic structure of complex disease [45, 46]. We used two different statistics, integrated haplotype score (iHS) [47] and Tajima's D, to measure the recent natural selection exerting on each gene category. iHS measured the increasing level of linkage disequilibrium surrounding a positively selected allele compared with the background allele at the same position [47]. The iHS of a gene was calculated using the proportion of SNPs with |iHS| > 2 in the gene. We found that iHS of MC genes and complex genes were very similar (average iHS = 0.075 and 0.077, respectively; Fig. 5.6a), and have been subjected to recent stronger positive selection compared with the other three categories of genes (Fig. 5.6a; $P < 0.02$,

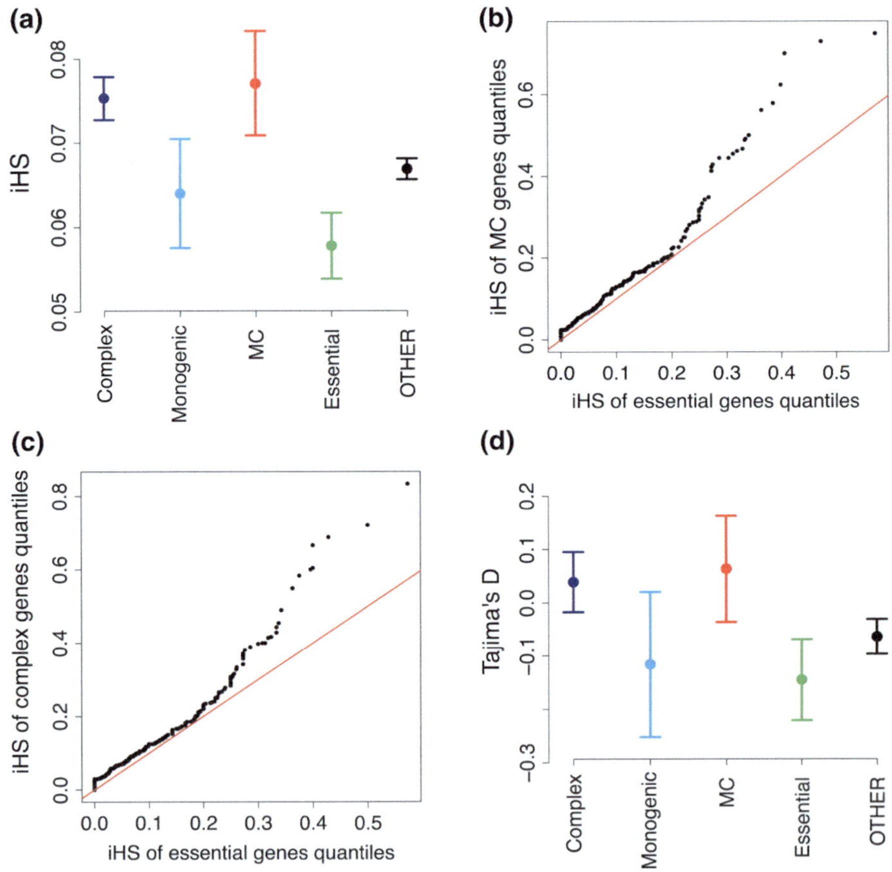

Fig. 5.6 Natural selection estimated from polymorphism data. **a** Comparison of iHS based on CEU in HapMap. **b** Q–Q plot of iHS between MC genes and essential genes. *Red line* shows null distribution that the values of MC genes are the same as those of essential genes. **c** Q–Q plot of iHS between complex genes and essential genes. *Red line* shows null distribution that the values of complex genes are the same as those of essential genes. **d** Comparison of Tajima's D based on European-American in Applera dataset. *Error bars* represent the standard error of the mean. Adapted from Jin et al. [60]

Wilcoxon test). The Q–Q plot of iHS between MC genes and essential genes showed that almost all points were above the null distribution (Fig. 5.6b), indicating that MC genes were under systematically stronger positive selection than essential genes. The extreme deviation of iHS at the tail of the distribution suggested that a fraction of MC genes has been subjected to especially stronger positive selection. Positive selection acting on complex genes was essentially the same as those on MC genes (Fig. 5.6c).

We calculated Tajima's D to examine natural selection in re-sequencing data (Applera dataset and NIEHS SNPs). The objective of Tajima's D is to distinguish DNA sequence evolving neutrally from the one evolving under a non-random process, including natural selection and demographic events [48]. Genes with extremely high or low Tajima's D should be the footprints of natural selection since human genes have experienced the same demographic history. Firstly, we calculated Tajima's D of each gene in European-American based on Applera dataset. We also found that Tajima's D of MC genes and complex genes were very similar (Fig. 5.6d). However, pair-wise comparisons of Tajima's D among all gene categories were not significantly different, except that between complex and essential genes showed significant difference without multiple comparisons correction ($P = 0.029$, Wilcoxon test) (Fig. 5.6d). These observations supported a previous report saying that some complex disease genes included targets of recent positive selection [20]. However, our much larger sets of gene showed that these differentiations were not as pronounced as that in the study of Blekhman et al. [20]. Although most of the MC and complex genes may have been subjected to recent positive selection compared with essential and monogenic genes, only a small fraction were under especially strong positive selection. We found that both MC and complex genes were significantly enriched in genes subjected to strong positive selection collected by Sabeti et al. [49] (Fig. 5.7).

5.8 Relative Evolutionary Rate Changed at Different Timescales

Elucidating the evolution of disease genes may shed light on the understanding of the origin of human diseases [11, 21], and for this goal, disease genes have been compared with non-disease genes in various studies [12–16, 50]. Evolutions of complex disease genes and monogenic disease genes also have been compared [19–21]. Although it has been shown consistently that disease genes are less conserved than housekeeping/essential genes, conflicting results do exist [12, 13, 17–21]. The contradictory observations may be rooted in the difference of timescales among these studies. To circumvent this problem, we performed a comparative analysis of the evolutionary rates at two main timescales based on nine genome-wide pair-wise divergence data.

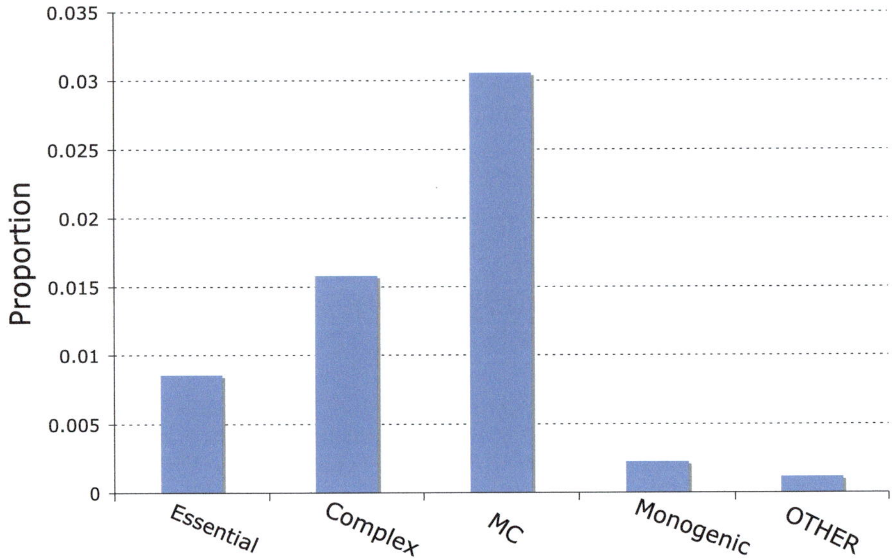

Fig. 5.7 Proportions of positive selection candidates in each gene category. Positive selection candidates collected by Sabeti et al. [49]

Firstly, we compared the evolutionary rate (d_N/d_S) of each gene since emergence of primates, which were calculated using the orthologous gene pairs between human and other primates. The results obtained from human–chimpanzee orthologous gene pairs showed that d_N/d_S of essential genes was the lowest while that of OTHER genes was the highest (Fig. 5.8a). Evolutionary rate of each kind of disease gene is significantly lower than that of OTHER genes ($P < 0.016$; Wilcoxon test), meanwhile significantly higher than that of essential genes ($P < 1.93 \times 10^{-5}$; Wilcoxon test), indicating that disease genes are more conserved than OTHER genes but evolve faster than essential genes [12]. Further analysis showed that the evolutionary rate of MC genes was 0.26 ± 0.013 (mean $\pm$ SEM), which was the lowest among the three disease gene categories (although not significantly different). Interestingly, genes with $d_N/d_S > 1$ existed in each gene category (Fig. 5.8b), which indicated that some genes have been subjected to positive selection since human and chimpanzee split. MC genes and essential genes had the highest proportion of genes with $d_N/d_S < 1$, which suggest that both of them had high proportion of genes under purifying selection. The major results are not qualitatively changed when either human–orangutan or human–rhesus macaque data were investigated, although the evolutionary rate of each pair-wise orthologous gene is different.

We further examined the evolutionary rates of disease genes in a larger timescale using six non-primate mammalians. The results obtained from human–mouse orthologous gene pairs showed that essential genes were still the most conserved (d_N/d_S is the lowest) among all gene categories (Fig. 5.8c). However, we found that not all kinds of disease genes were more conserved than OTHER

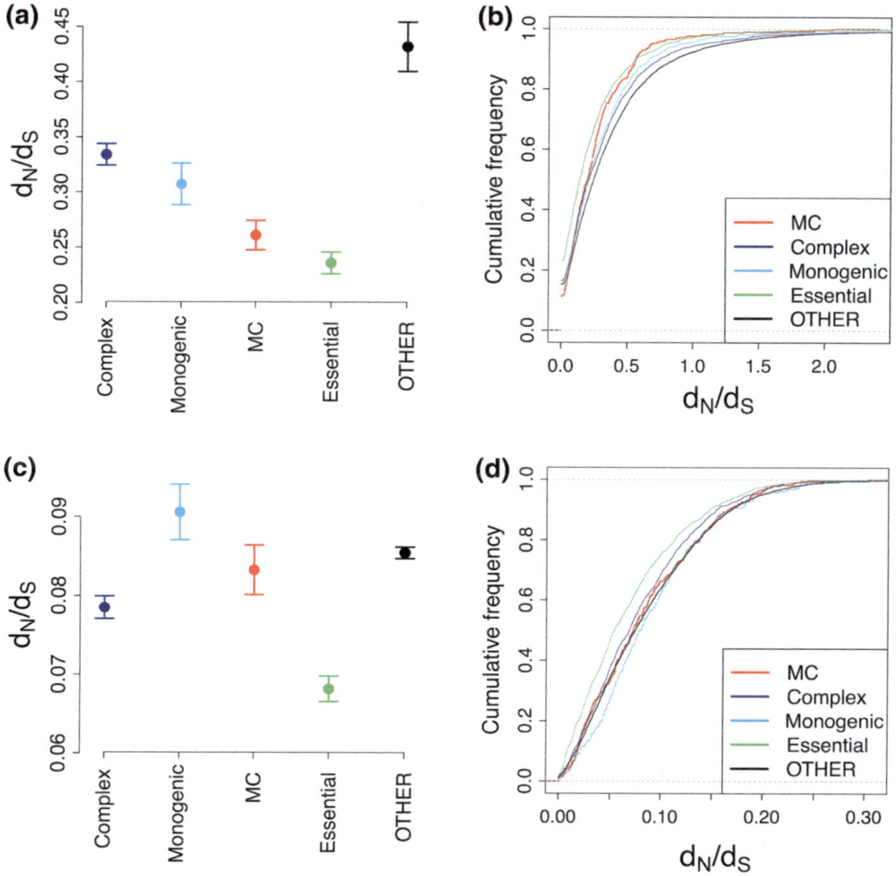

Fig. 5.8 Evolutionary rates of each gene category estimated from divergence data. *Error bars* represent the standard error of the mean. **a** Comparison of d_N/d_S based on human–chimpanzee orthologous gene pairs. **b** Cumulative distribution of d_N/d_S based on human–chimpanzee orthologous gene pairs. **c** Comparison of d_N/d_S based on human–mouse orthologous gene pairs. **d** Cumulative distribution of d_N/d_S based on human–mouse orthologous gene pairs. Adapted from Jin et al. [60]

genes (e.g., monogenic genes evolved even faster than OTHER genes), which was different from that observed in primate lineage. Interestingly, there is no gene with $d_N/d_S > 1$ in all gene categories (Fig. 5.8d), which indicates that all genes are under purifying selection in long timescale, while positive selection is a temporary event for a specific gene. The results by comparing with other non-primate mammalians such as rat, cow, pig, rabbit, and panda were similar to that by mouse.

Our observations still held even when we used only the orthologous gene pairs with >90 % identity. We proposed a hypothesis to explain the changed relative evolutionary rate at different timescales: Human disease genes might play very important roles in primate lineage, which made these genes more conserved than OTHER genes. However, some of these disease genes might have different functions in non-primate mammalians compared with their orthologous genes in human, which makes these genes no more conserved than OTHER genes. In short, genes underlying human genetic disorders might play different roles in non-primate mammalians. Studies trying to identify the functions and pathways of human disease gene or to explore the etiology of human diseases in animal model such as mouse and rat should be cautious, because the conclusions in animal model may no longer hold in human. We also found that some genes were under positive selection in primate lineage, but did not find any gene subjected to positive selection when compared with the non-primate mammalians. These observations indicate that the positive selection is only a temporary phenomenon for a specified gene, while purifying selection lasts in the long evolutionary history.

5.9 Over-representation Analysis with CNV Regions

Copy number variations (CNVs) could significantly disrupt the functions of a gene, especially when CNVs are located in genic regions [51, 52], on the other hand, the fate of a CNV in a population (loss or fixation) is also significantly impacted by its effect on the gene function. In particular, a CNV cannot disperse and maintain in a population if it impairs gene functions that lead to lethality. For example, it has been shown that CNVs are relatively scarce in genic regions compared with nongenic regions [52]. To investigate the effect of CNVs on each gene category, we conducted a series of enrichment analyses based on the validated CNVs by Conrad et al. [52]. Interestingly, complex genes are significantly enriched in CNV regions ($P = 5.16 \times 10^{-6}$, Fisher's exact test) compared with non-complex genes (Table 5.1), which might implicate that genes located in CNV regions are more likely to cause complex diseases. This enrichment is more significant when only the genes located in common deletion CNVs were considered ($P = 2.59 \times 10^{-16}$,

Table 5.1 Enrichment/impoverishment of genes for each gene category in validated CNVs

CNV / Genes	All CNVs	Gen.Dups	Gen.Dels	Gen.Multi	Common CNVs	Common Dups	Common Dels	Rare CNVs	Rare Dups	Rare Dels
Essential	8.15×10^{-4}	1.13×10^{-7}	0.722	7.22×10^{-6}	1.27×10^{-4}	9.47×10^{-7}	0.213	0.046	2.31×10^{-5}	0.80
Complex	5.16×10^{-6}	0.965	2.31×10^{-15}	0.339	3.79×10^{-7}	0.91	2.59×10^{-16}	1.04×10^{-4}	0.616	6.63×10^{-9}
MC	0.743	0.223	0.614	0.563	0.599	0.337	0.819	0.934	0.122	0.344
Monogenic	0.025	0.041	0.358	4.26×10^{-4}	0.021	0.049	0.26	0.241	0.072	0.82

P-values were calculated using Fisher's exact test based on the number of genes located in CNVs or not. The *red* implicates significant enrichment and *green* implicates significant impoverishment. 'Gen' = genotyped; 'Common' = MAF $\geq$ 10 %, 'Rare' = MAF < 10 %. Adapted from Jin et al. [60]

Fisher's exact test) (Table 5.1). In contrast, essential genes showed a significant impoverishment of genes located in CNV regions ($P = 8.15 \times 10^{-4}$, Fisher's exact test) compared with non-essential genes (Table 5.1). Monogenic genes were also found to show impoverishment of genes located in CNV regions ($P = 0.025$, Fisher's exact test), and were more significant for these genes located in multiallelic CNVs ($P = 4.26 \times 10^{-4}$, Fisher's exact test). As for MC genes, no significant associations with CNV regions were observed. The results based on all detected CNVs by Conrad et al. [52] were similar to that of the validated CNVs.

Assuming that original mutation rates of CNVs may not favor any gene category, why were essential genes and monogenic genes significantly impoverished in CNV regions, while complex genes were significantly enriched in CNV regions? One reasonable explanation is that CNV mutations happened on essential genes and monogenic genes might be lethal or seriously affect the individuals, and the strong purifying selections may finally lead to the disappearance of these CNVs, which leads to the under-representation of essential genes and monogenic genes in CNV regions. Many CNV mutations that happened on complex genes may not lead to very serious illnesses or phenotypes at early age and may subject to positive selection in some cases, which finally leads to the over-representation of complex genes in CNV regions. Our analyses indicate that carefully examining CNVs around SNPs associated with complex diseases may finally reveal their pathology since complex disease genes are likely to be caused by CNVs. In this way, MC genes must have been subjected to both positive and purifying selection due to their double identities, which is consistent with our analyses based on polymorphism data and divergence data.

5.10 Distinct Functional Characteristics of Different Genes Categories

Based on DAVID, disease mutation was significantly enriched in MC genes, monogenic genes, and essential genes ($P < 1.65 \times 10^{-122}$, FDR). Most importantly, essential genes, MC genes, complex genes, and monogenic genes had different functional characteristics. Gene ontology (GO) analyses revealed that genes involved in embryonic development and basic cell activities were significantly enriched in essential genes. Among the 1,520 essential genes, 179 were involved in embryonic development ending in birth or egg hatching ($P = 5.50 \times 10^{-81}$, FDR), 166 in embryonic morphogenesis ($P = 1.92 \times 10^{-69}$, FDR), 363 in DNA binding ($P = 1.46 \times 10^{-61}$, FDR), and 469 in the regulation of transcription ($P = 4.92 \times 10^{-31}$, FDR). The genes related to homeostatic processes and body fluid were significantly enriched in MC genes. For example, 117 genes were involved in homeostatic processes ($P = 9.72 \times 10^{-38}$, FDR), 98 in chemical homeostasis ($P = 3.44 \times 10^{-35}$, FDR), 44 in the regulation of body fluid levels ($P = 2.57 \times 10^{-24}$, FDR), and 49 in blood circulation ($P = 8.73 \times 10^{-24}$, FDR). For monogenic genes, the most significantly enriched clusters were visual perception ($P = 4.35 \times 10^{-17}$, FDR) and sensory perception of light stimulus ($P = 4.35 \times 10^{-17}$, FDR), followed by ectoderm development ($P = 5.40 \times 10^{-9}$, FDR) and epidermis development ($P = 2.63 \times 10^{-8}$, FDR). Complex genes were significantly enriched for genes involved in regulation and defense system, such as regulation of cell proliferation ($P = 9.83 \times 10^{-33}$, FDR), response to wounding ($P = 1.11 \times 10^{-30}$, FDR), defense response ($P = 3.05 \times 10^{-30}$, FDR), and immune response ($P = 1.89 \times 10^{-28}$, FDR).

5.11 Summary and Genetic Implication

For the past century, monogenic diseases and complex diseases have been considered as distinct groups at the opposite ends of the spectrum of human diseases. Although it has been reported some genes are associated with both monogenic and complex diseases [2, 23, 25, 53], which have not been systematically compiled. For example, normal genetic variability that contributes to the risk of sporadic diseases has been reported located at the genes (such as *MAPT* and *APP*) encoding the deposited protein in monogenic diseases [53–55]. We would like to highlight our observation that monogenic disease genes are more likely to be involved in complex diseases, which is a strong evidence to justify that genes involved in related monogenic disease should be selected as candidate genes for candidate gene association studies. On the other hand, many gene category-specific characteristics could not be illustrated if MC genes were not separated from the other disease gene categories. For examples, we could not find the significant impoverishment of monogenic disease genes in CNV regions without separating the MC genes from monogenic genes. Furthermore, we expected the number of MC genes to be increased considerably in the future as the genes underlying monogenic disease and complex diseases are both accumulating. Genome-wide association studies always provide new insights into gene functions, which would help us review the pathogenic of monogenic diseases. By studying monogenic diseases caused by MC genes, functions and networks of these genes in complex disease might be revealed. In this way, progresses in both monogenic disease and complex disease will mutually benefit each other in the future.

In order to gain global insights and perspectives on these characteristics, potential relationships among these characteristics are discussed below. Firstly, MC genes were significantly associated with more diseases and phenotypes than monogenic genes and complex genes on average. Our analyses also showed that there were more interacting proteins for MC genes than that for the monogenic genes and complex genes on average, which may indicate that genes with more interacting proteins tend to be involved in more diseases and phenotypes. Secondly, essential genes have the highest average connectivity in protein–protein interaction network and are the most conserved among the five gene categories, supporting previous studies that genes with more interacting proteins tend to evolve more slowly than those with fewer ones [56, 57]. Thirdly, MC genes had the longest CDS and highest transcript count among the five gene categories, which may indicate why they could interact with more proteins than the other two kinds of disease genes. Although expression level of MC genes was not significantly higher than that of monogenic and complex genes, its tissue specificity was the highest among all gene categories. In addition, our analysis also showed that different gene categories showed different functional characteristics. In short, MC genes, encoding the longest proteins, having the highest tissue specificity and the highest transcript count among all gene categories, tend to evolve in more complex protein–protein interaction network and thus involved in much more genetic diseases.

Genetic adaptation to environmental challenge such as pathogens, climate, and diet has undoubtedly been important to human survival [58, 59]. In this study, we not only analyzed the recent natural selection using human polymorphism data, but also investigated the selection in our ancestors using genome-wide divergence data. We are interested in whether these selections are consistent at different timescales such as inter-species and intra-species. Firstly, based on haplotype data, our analyses showed that both complex and MC genes were under stronger recent positive selection compared with monogenic and essential genes, which reflected the signatures of natural selection in recent thousands of years [49]. It is well known that genes under positive selection evolve faster than those under purifying selection; and genes with more interacting proteins tend to evolve slower [56, 57]. So we predicted that essential genes are the most conserved, and MC genes are more conserved than monogenic and complex genes in the inter-species level. Actually, our observations based on divergence data in primate lineage are consistent with the predictions. In a nutshell, our analysis showed that essential genes are the most conserved in each evolutionary timescale. Monogenic genes were under purifying selection thus conserved in primate lineage, but they evolved the fastest (or the second fastest) among all gene categories in non-primate mammalian scales. These phenomena are consistent with the results of previous reports, i.e., genes with less interacting proteins evolved faster than those with more [56, 57]. MC and complex genes demonstrated complex selection patterns across timescales, which might be caused by the recent positive selections acting on them. In particular, over-representation analysis of CNVs in different gene categories also suggested that MC genes have been subjected to both strong purifying and positive selection.

References

1. Antonarakis SE, Chakravarti A, Cohen JC, Hardy J. Mendelian disorders and multifactorial traits: the big divide or one for all? Nat Rev Genet. 2010;11:380–4.
2. Sidransky E. Heterozygosity for a Mendelian disorder as a risk factor for complex disease. Clin Genet. 2006;70:275–82.
3. Botstein D, Risch N. Discovering genotypes underlying human phenotypes: past successes for mendelian disease, future approaches for complex disease. Nat Genet. 2003;33(Suppl):228–37.

4. Altshuler D, Daly MJ, Lander ES. Genetic mapping in human disease. Science. 2008;322:881–8.
5. Antonarakis SE, Beckmann JS. Mendelian disorders deserve more attention. Nat Rev Genet. 2006;7:277–82.
6. McKusick VA. Mendelian inheritance in man and its online version, OMIM. Am J Hum Genet. 2007;80:588–604.
7. Amberger J, Bocchini CA, Scott AF, Hamosh A. McKusick's online mendelian inheritance in man (OMIM). Nucleic Acids Res. 2009;37:D793–6.
8. Hindorff LA, et al. Potential etiologic and functional implications of genome-wide association loci for human diseases and traits. Proc Natl Acad Sci USA. 2009;106:9362–7.
9. McCarthy MI, et al. Genome-wide association studies for complex traits: consensus, uncertainty and challenges. Nat Rev Genet. 2008;9:356–69.
10. Frazer KA, Murray SS, Schork NJ, Topol EJ. Human genetic variation and its contribution to complex traits. Nat Rev Genet. 2009;10:241–51.
11. Keller MC, Miller G. Resolving the paradox of common, harmful, heritable mental disorders: which evolutionary genetic models work best? Behav Brain Sci. 2006;29:385–404; discussion 405–52.
12. Tu Z, et al. Further understanding human disease genes by comparing with housekeeping genes and other genes. BMC Genomics. 2006;7:31.
13. Smith NG, Eyre-Walker A. Human disease genes: patterns and predictions. Gene. 2003;318:169–75.
14. Kondrashov FA, Ogurtsov AY, Kondrashov AS. Bioinformatical assay of human gene morbidity. Nucleic Acids Res. 2004;32:1731–7.
15. Winter EE, Goodstadt L, Ponting CP. Elevated rates of protein secretion, evolution, and disease among tissue-specific genes. Genome Res. 2004;14:54–61.
16. Bustamante CD, et al. Natural selection on protein-coding genes in the human genome. Nature. 2005;437:1153–7.
17. Huang H, et al. Evolutionary conservation and selection of human disease gene orthologs in the rat and mouse genomes. Genome Biol. 2004;5:R47.
18. Lopez-Bigas N, Ouzounis CA. Genome-wide identification of genes likely to be involved in human genetic disease. Nucleic Acids Res. 2004;32:3108–14.
19. Thomas PD, Kejariwal A. Coding single-nucleotide polymorphisms associated with complex vs. Mendelian disease: evolutionary evidence for differences in molecular effects. Proc Natl Acad Sci USA. 2004;101:15398–403.
20. Blekhman R, et al. Natural selection on genes that underlie human disease susceptibility. Curr Biol. 2008;18:883–9.
21. Podder S, Ghosh TC. Exploring the differences in evolutionary rates between monogenic and polygenic disease genes in human. Mol Biol Evol. 2010;27:934–41.
22. Cooper DN, Mort M. Do inherited disease genes have distinguishing functional characteristics? Genetic Testing Molecular Biomarkers. 2010;14:289–91.
23. Antoniou AC, Chenevix-Trench G. Common genetic variants and cancer risk in Mendelian cancer syndromes. Curr Opin Genet Dev. 2010;20:299–307.
24. Hinney A, Vogel CI, Hebebrand J. From monogenic to polygenic obesity: recent advances. Eur Child Adolesc Psychiatry. 2010;19:297–310.
25. Bultron G, et al. The risk of Parkinson's disease in type 1 Gaucher disease. J Inherit Metab Dis. 2010;33:167–73.
26. Becker KG, Barnes KC, Bright TJ, Wang SA. The genetic association database. Nat Genet. 2004;36:431–2.
27. Zhang Y, et al. Systematic analysis, comparison, and integration of disease based human genetic association data and mouse genetic phenotypic information. BMC Med Genomics. 2010;3:1.
28. Keshava Prasad TS, et al. Human protein reference database–2009 update. Nucleic Acids Res. 2009;37:D767–72.

29. Feldman I, Rzhetsky A, Vitkup D. Network properties of genes harboring inherited disease mutations. Proc Natl Acad Sci USA. 2008;105:4323–8.
30. Liao BY, Scott NM, Zhang J. Impacts of gene essentiality, expression pattern, and gene compactness on the evolutionary rate of mammalian proteins. Mol Biol Evol. 2006;23:2072–80.
31. Goh KI, et al. The human disease network. Proc Natl Acad Sci USA. 2007;104:8685–90.
32. Gandhi TK, et al. Analysis of the human protein interactome and comparison with yeast, worm and fly interaction datasets. Nat Genet. 2006;38:285–93.
33. Lopez-Bigas N, Audit B, Ouzounis C, Parra G, Guigo R. Are splicing mutations the most frequent cause of hereditary disease? Febs Lett. 2005;579:1900–3.
34. Dermitzakis ET. From gene expression to disease risk. Nat Genet. 2008;40:492–3.
35. Cookson W, Liang L, Abecasis G, Moffatt M, Lathrop M. Mapping complex disease traits with global gene expression. Nat Rev Genet. 2009;10:184–94.
36. Kilpinen S, et al. Systematic bioinformatic analysis of expression levels of 17,330 human genes across 9,783 samples from 175 types of healthy and pathological tissues. Genome Biol. 2008;9:R139.
37. Yanai I, et al. Genome-wide midrange transcription profiles reveal expression level relationships in human tissue specification. Bioinformatics. 2005;21:650–9.
38. Chen R, et al. FitSNPs: highly differentially expressed genes are more likely to have variants associated with disease. Genome Biol. 2008;9:R170.
39. Lage K, et al. A large-scale analysis of tissue-specific pathology and gene expression of human disease genes and complexes. Proc Natl Acad Sci USA. 2008;105:20870–5.
40. Nicolae DL, et al. Trait-Associated SNPs Are More Likely to Be eQTLs: annotation to enhance discovery from GWAS. Plos Genet. 2010;6.
41. Zhang L, Li WH. Mammalian housekeeping genes evolve more slowly than tissue-specific genes. Mol Biol Evol. 2004;21:236–9.
42. Eisenberg E, Levanon EY. Human housekeeping genes are compact. Trends in Genetics. 2003;19:362–5.
43. Wilson AC, Carlson SS, White TJ. Biochemical evolution. Annu Rev Biochem. 1977;46:573–639.
44. Hirsh AE, Fraser HB. Protein dispensability and rate of evolution. Nature. 2001;411:1046–9.
45. Kryukov GV, Pennacchio LA, Sunyaev SR. Most rare missense alleles are deleterious in humans: implications for complex disease and association studies. American Journal of Human Genetics. 2007;80:727–739
46. Di Rienzo A, Hudson RR. An evolutionary framework for common diseases: the ancestral-susceptibility model. Trends in Genetics. 2005;21:596–601
47. Voight BF, Kudaravalli S, Wen XQ, Pritchard JK. A map of recent positive selection in the human genome. Plos Biol. 2006;4:446–58.
48. Tajima F. Statistical method for testing the neutral mutation hypothesis by DNA polymorphism. Genetics. 1989;123:585–95.
49. Sabeti PC, et al. Positive natural selection in the human lineage. Science. 2006;312:1614–20.
50. Cai JJ, Borenstein E, Chen R, Petrov DA. Similarly strong purifying selection acts on human disease genes of all evolutionary ages. Genome Biol Evol. 2009;2009:131–44.
51. Hurles ME, Dermitzakis ET, Tyler-Smith C. The functional impact of structural variation in humans. Trends Genet. 2008;24:238–45.
52. Conrad DF, et al. Origins and functional impact of copy number variation in the human genome. Nature. 2010;464:704–12.
53. Singleton A, Myers A, Hardy J. The law of mass action applied to neurodegenerative disease: a hypothesis concerning the etiology and pathogenesis of complex diseases. Hum Mol Genet 2004;13 spec No 1:R123–6.
54. Hutton M, et al. Association of missense and 5′-splice-site mutations in tau with the inherited dementia FTDP-17. Nature. 1998;393:702–5.
55. Lambert JC, et al. Genome-wide association study identifies variants at CLU and CR1 associated with Alzheimer's disease. Nat Genet. 2009;41:1094–9.

56. Fraser HB, Hirsh AE, Steinmetz LM, Scharfe C, Feldman MW. Evolutionary rate in the protein interaction network. Science. 2002;296:750–2.
57. Fraser HB, Wall DP, Hirsh AE. A simple dependence between protein evolution rate and the number of protein-protein interactions. BMC Evol Biol. 2003;3:11.
58. Balaresque PL, Ballereau SJ, Jobling MA. Challenges in human genetic diversity: demographic history and adaptation. Hum Mol Genet. 2007;16(R2):R134–9.
59. Oleksyk TK, Smith MW, O'Brien SJ. Genome-wide scans for footprints of natural selection. Philos Trans R Soc Lond B Biol Sci. 2010;365:185–205.
60. Jin W, Qin P, Lou H, Jin L, Xu S. A systematic characterization of genes underlying both complex and Mendelian diseases. Hum Mol Genet. 2012;21:1611–24.

Chapter 6
Materials and Methods

Abstract Total 1,890 unrelated African Americans and 458 unrelated Mexicans, as well as thousands of unrelated individuals from their potential parental populations were collected. Each of these individuals has been genotyped by one or more genome-wide microarrays and hundreds of thousands of SNPs for each individual were obtained. The population structure and population admixture has been analyzed by various approaches, such as STRUCTURE, FRAPPE, PCA and HAMPIX. On the other hand, three admixture models (HI model, GA model and CGF mode) have been introduced and simulations have been conducted to fit the empirical data. Earth mover's distance has been used to quantitatively measure the difference of LACS distribution between simulated and empirical data. Finally, human polymorphic data, species divergence data, copy number variations (CNV) data, gene expression data and protein interaction data were collected to characterize the disease genes.

Keywords Genotyping data · Population structure analysis · Admixture models · Computer simulation · Earth mover's distance (EMD) · Copy number variation (CNV)

6.1 Population Admixture Models

In reality, population admixture processes are either unknown or too complex to study directly. Here, we explored population admixture dynamics by examining distribution of length of ancestral chromosomal segments (LACS) in three typical admixture models which can summarize most of the possible scenarios: Hybrid isolation (HI) model [1], gradual admixture (GA) model [2], and continuous-gene-flow (CGF; an extension of GA model) model [3, 4]. The genetic structure and linkage disequilibrium pattern of the admixed population under these models have

W. Jin, *Admixture Dynamics, Natural Selection and Diseases in Admixed Populations*, Springer Theses, DOI 10.1007/978-94-017-7408-6_6

been investigated systematically in several studies [2–5]. In the three admixture models, m and $1 - m$ represent the genetic contribution of parental populations, pop1 and pop2, to the admixed population. T represents the final number of generations since the first admixture, while t represents the number of generations the admixed population has experienced, ranging from 1 to T. In the HI model, admixture occurs only in the first generation without further genetic contribution from parental populations. We assumed that the admixed population experiences recombination and genetic drift only in the following generations. In the GA model, admixture occurs gradually and the parental populations contribute gene flow to the admixed population in each generation. The rates of gene flow from pop1 and pop2 to the admixed population in each generation are m/T and $(1 - m)/T$, respectively. This ensures that the genetic contribution of a given parental population to the admixed population is constant per generation. In the CGF model, one parental population (genetic donor) continuously contributes a constant but reduced rate of gene flow (α) to the admixed population per generation, so the cumulative genetic contribution from both parental populations is equal to that under HI model and GA model. The gene flow that the admixed population receives from the genetic donor per generation is calculated using $\alpha = 1 - (m)^{1/T}$.

6.2 Data Simulation and Comparison with Theoretical LACS Distribution

A forward-time simulation program was developed based on the three aforementioned admixture models. We used phased genotype data of YRI (Yoruba in Ibadan, Nigeria) and CEU (Utah residents with northern and western European ancestry from the CEPH collection) from HapMap as the initial statuses of the parental populations [6]. We first sampled the haploid chromosomes of YRI and CEU according to their estimated contribution, respectively. Then we combined each pair of haploid chromosomes from the two parental populations to construct a diploid admixed individual. We modeled the processes of genetic drift and recombination under a Wright–Fisher neutral model [7]. In the simulation, recombination was introduced according to the genetic map adapted from HapMap (release #22; with 3,540 cM in total on the 22 autosomes) [6], and mutation was ignored given the short population history studied. The effective population size (N_e) of each population was set at 10,000. We labeled the genotypes from different parental populations when no empirical data were integrated for analysis. In the CGF model, for a given specific admixture proportion, the parental population can either serve as genetic donor or as genetic recipient thus experiencing different histories. In this way, the genetic donor in CGF model was referred to as CGFD, while the genetic recipient in CGF model was referred to as CGFR. The number of generations for each model was set to be 10, 20, 50, and 100, respectively. Extensive simulations were performed to explore the influence of admixture dynamics on the distribution of LACS by modifying various parameters such as proportions of ancestry contribution and N_e. Ancestral origin of each locus in the haplotype was labeled to track ancestral chromosomal segments in the simulated admixed population, which allowed us to directly compare the simulated LACS distribution with the theoretical distribution of LACS that was calculated using the formula deduced in this study.

6.3 Simulation of Susceptibility Locus in Admixed Population and Admixture Mapping

To elucidate the influences of LACS distribution on admixture mapping in HI and GA models, we simulated a dataset for a systematic comparative analysis. Based on the aforementioned methods for the simulation of admixed population, we randomly sampled the simulated admixed individuals as controls. Cases were simulated by random sampling of haploid chromosomes from admixed individuals, but restricted genetic contribution from the parental population particularly in the susceptibility locus.

More specifically, we treated the haploid chromosomes of YRI and CEU as the initial status of the two parental populations. Then we combined each pair of haploid chromosomes from parental populations to construct the initial diploid admixed individual. The genetic contribution of the given parental population to the admixed population (θ) was set at 20 %, which was similar to the genetic contribution of European to African Americans. The number of generations since the initial population admixture (λ) was set as 20, which is an approximation of the generation of population admixture in the New World. Finally, we set the sample size of case and control to 2000 and assumed an increased disease risk of 2 for the susceptibility locus containing an allele from the given parental population in both HI and GA models, relative to the alleles that did not come from the other parental population. We compared the signatures of association in HI and GA models based on case-only and case-control approaches.

6.4 Genetic Data of African Americans

The genotypic data were obtained from International HapMap Project (HapMap; http://www.hapmap.org), Human Genome Diversity Project (HGDP; http://www.cephb.fr/en/hgdp), and Illumina iControlDB (http://www.illumina.com), respectively. The combined dataset includes the genotypic data of 588 HapMap samples [6], 300 HGDP samples (156 indigenous European, 102 indigenous African, and 42 Amerindian) [8], 2,161 African Americans, and 3,294 Caucasians from iControlDB genotyped by Illumina 550K Beadarray. The samples from iControlDB have passed Illumina's rigorous quality control and have been used as controls in five genome-wide association studies. In detail, the 588

HapMap samples include 87 ASW (individuals of African ancestry from the southwest U.S.), 167 YRI (Yoruba in Ibadan, Nigeria), 165 CEU (Utah residents with northern and western European ancestry from the CEPH collection), 85 CHD (Han Chinese from metropolitan Denver, CO, USA), and 84 CHB (Han Chinese from Beijing). The 102 HGDP indigenous Africans from seven different geographic or linguistic groups (including 21 Yoruba from West Central Africa, 22 Mandenka from West Africa, 11 Bantu speakers from Kenya, 8 Bantu speakers from Southwestern Africa, 22 Biaka Pygmy and 13 Mbuti Pygmy from Central Africa, and 5 San from Southern African) [8] represent the potential African ancestry of African American. The 156 HGDP indigenous Europeans from eight different geographic and linguistic groups (including 17 Adygei from Russia, 24 Basque from France, 28 French in France, 12 Italian from North Italy, 15 Orcadian from Orkney Islands, 25 Russian in European, 28 Sardinian in Sardinia island, and 7 Tuscan from Italy) [8] represent the potential European ancestry of African American.

ASW from HapMap and African Americans from iControlDB merged into one African Americans population, and Yoruba from HGDP was collected from Nigeria and therefore merged with YRI from HapMap in subsequent analyses. The following samples were removed from the dataset: (1) the relatives based on sample information or PLINK [9] result (IBD > 0.2), (2) individuals identified as outliers of each population (except African Americans) based on the top ten principal components in PCA analysis (SD > 6), (3) African Americans individuals with >2 % Native-American/East-Asian, (4) African Americans individuals with >99 % European contribution, which are likely to be descendents of individuals of European ancestry, and (5) African Americans individuals with >99 % African contribution, in which recent African immigrants could not be practically identified [10].

These filtered samples described above were merged and the SNPs sharing reference SNP ID (rs) and vendor-specified strands were kept in combined data. Then, the dataset was further filtered for individuals with >10 % missing genotypes and SNPs with >10 % missing data, as well as Hardy–Weinberg disequilibrium ($P < 2 \times 10^{-6}$) within each population except African Americans. The final dataset comprised of 503,694 SNPs (491,526 autosomal SNPs) shared by 5,210 individuals from 21 population groups, with total genotyping call rate 99.74 %.

Overall, 1,890 unrelated African Americans samples with ignorable ancestry outside of Africa and Europe were studied, among which 1,838 individuals were downloaded from iControlDB, and another 52 individuals were from the International HapMap Project [6]. In addition, 113 YRI from HapMap [6] and 102 indigenous Africans in seven different groups collected from HGDP [8] were merged, representing the extant Africans, while 113 CEU from HapMap [6] were merged with 156 Europeans in eight different groups collected from HGDP, representing the extant Europeans. Total 2,648 CAU-GWAS samples from iControlDB were taken as another representation of the extant Europeans. In addition, 84 CHB and 85 CHD represented populations from East Asia, and 24 pure Amerindians from HGDP represented Native American.

6.5 Genetic Data of Mexicans

Mexican-Mestizos and Mexican-Americans are recently admixed populations mainly composed of Amerindian and European with similar history, both of which were referred to as Mexicans for the convenience of presentation in this study. Firstly, 300 unrelated self-identified Mexican-Mestizo individuals and 239 unrelated individuals from their putative parental populations in MGDP were downloaded from the INMEGEN website [11]. These individuals were genotyped using Affymetrix 100K SNP array and 99,953 SNPs passed through quality control. The population structure of these Mestizos has been systemically analyzed in the study by Silva-Zolezzi [11]. Mexican-American samples were obtained from two datasets: The first dataset contained 100 Mexicans from the Coriell HD100MEX panel genotyped on the Affymetrix SNP 6.0 GeneChip. The second dataset contained 58 unrelated Mexicans from HapMap3 genotyped on both Affymetrix SNP 6.0 GeneChip and Illumina 1M Beadarray. The two datasets of Mexican-Americans were merged because both of them were collected in Los Angeles and genotyped on Affymetrix SNP 6.0 GeneChip. Overall, 767,454 autosomal SNPs on the 158 Mexican-Americans were kept after quality control.

Overall, 458 unrelated Mexicans including 300 Mexican-Mestizos and 158 Mexican-Americans were collected. Since both historical records and genetic evidences indicate that African populations have contributed only mildly to the Mexican gene pool, 45 samples with pronounced African ancestral components were filtered out. Thus we simplified the following analysis to a two-way admixture between European and Amerindian populations. After filtering out SNPs with >10 % missing genotypes, we had 36,000 autosomal SNPs shared by the remaining 652 samples (413 Mexicans and 239 samples from parental populations). PCA was performed at the individual level using all 36,000 autosomal SNPs shared by the 652 samples.

6.6 Linkage Disequilibrium Correction and PCA Analysis

Principle component analysis (PCA) is a statistical method widely used to identify population genetic variation across geographical location and ethnic background. It is well known that the uncorrected linkage disequilibrium (LD) would seriously distort the results of PCA and thus bias the population structure [12]. In order to reduce the LD between markers, we used the parameter "-indep-pair-wise 50 5 0.5" in PLINK [9] to remove any SNP that had a pair-wise r^2 greater than 0.5 in window of 50 SNPs, shifted and recalculated every 5 SNPs. Based on these thinned markers, PCA was performed using *smartpca* in the EIGENSOFT package [12] under default setting with no outlier removal. In the PCA analysis, the first principal component (PC1) is the coordinate drawn in the multidimensional space so that the projections of the points (each point represents an individual) to the coordinate have the largest variance. The second principal component (PC2) is the coordinate drawn in the multidimensional space so that the projections of the points to the coordinate have the second largest variance, and so forth. Intuitively, PC1 and PC2 could reflect the two primary genetic differences between samples. In this study, only PC1 and PC2 were plotted in our PCA plot and have been carefully analyzed.

The projections of individuals onto existing axes not only identified the population structure, but also could identify admixture proportion of each admixed individual [13]. So we performed PCA on African American and its two parental populations (YRI and CEU), and inferred the proportion of each African American based on projection of samples on PC1.

6.7 Population Genetic Analysis and Inference of Ancestral Chromosomal Segments

The genetic contribution of the parental population to the admixed population was inferred using FRAPPE, which implements an expectation–maximization (EM) algorithm [14]. FRAPPE was run on all the SNPs available with 10,000 iterations by setting $K = 2$. We also ran STRUCTURE [15, 16], which implemented a model-based clustering method to infer population structure using genome-wide unlinked markers. STRUCTURE was run with 100,000 burn-ins and 100,000 iterations by setting admixture model assuming that the allele frequencies were correlated. Genetic difference between populations was measured using F_{ST} following Weir and Cockerham [17], which accounts for differences in the sample size in each population. The locus-specific F_{ST} between any two populations was also calculated using the same formula.

We chose HAPMIX to infer ancestral chromosomal segments in the admixed populations in this study as several previous studies using simulated data had reported that HAPMIX outperformed the other available methods and software [18, 19]. The haplotypes of parental populations, the inputs for HAPMIX, were either downloaded directly from HapMap website or phased using fastPHASE [20] when the phased data were unavailable. When we ran fastPHASE, population labels were used during the model fitting to enhance accuracy, and haplotypes from the HapMap YRI and CEU were included as known because they were phased in trios and were assumed to be highly accurate. We found a lot of extremely short ancestral chromosomal segments, which might have resulted from some uncertainty in statistical inference. We therefore, removed those very short ancestral chromosomal segments in order to improve the reliability of the distribution.

6.8 Earth Mover's Distance (EMD)

EMD was first introduced as an empirical way to measure color and texture similarities by Rubner et al. [21]. In computer sciences, it measures the dissimilarity or distance between two probability distributions across one region [22, 23]. Later, it has been demonstrated as conceptually equivalent to Wasserstein metric [24]. Intuitively, given two distributions, one can be seen as a mass of earth properly spread in space and the other as a collection of holes in the same space. Then the EMD calculates the least amount of work needed to fill the holes with the earth. In this way, EMD is the minimum cost of turning one pile into the other, where the cost is assumed to be the amount of dirt moved times the distance by which it is moved.

Computing EMD is essentially similar to the solution of the well-known transportation problem [22]. There are m suppliers, each with a given amount of goods, and n consumers, each with a given limited capacity. For each supplier-consumer pair, the cost of transportation correlated with the amount of good and distance. This can be formalized as a linear programming problem: Let $P = \{(p_1, w_{p_1}), \ldots, (p_m, w_{p_m})\}$ be the first signature with m clusters, where w_p is the cluster weight; $Q = \{(q_1, w_{q_1}), \ldots, (q_m, w_{q_m})\}$ is the second signature with n clusters; and the distance matrix $D = [d_{ij}]$ where d_{ij} is the distance between cluster p_i and q_j. The flow $F = [f_{ij}]$, with f_{ij} the flow between p_i and q_j, minimizes the overall cost

$$\mathrm{WORK}(P, Q, F) = \sum_{i=1}^{m} \sum_{j=1}^{n} f_{ij} d_{ij}$$

We would find the optimal flow F if the transportation problem is solved. Earth mover's distance is defined as the work normalized by the total flow:

$$\mathrm{EMD}(P, Q) = \frac{\sum_{i=1}^{m} \sum_{j=1}^{n} f_{ij} d_{ij}}{\sum_{i=1}^{m} \sum_{j=1}^{n} f_{ij}}$$

The normalization factor was introduced to avoid favoring smaller signatures in the case of partial matching [24].

6.9 Measurement of the Differences Between Two Distributions

In this study, statistical computing and graphics generating were mainly performed using R version 2.13 [25]. The 10-based logarithms of LACS were calculated by assuming log-normal distribution of LACS. The log-normal distribution was displayed using the probability density function, which describes the relative likelihood that a given random variable occurs at a given point. The probability density function is nonnegative everywhere, and its integral over the entire space is equal to 1. To assess the differences of distribution of LACS among different models and empirical data, Kolmogorov–Smirnov (K–S) tests [26] were performed. K–S test is a nonparametric test for the equality of continuous, one-dimensional probability distributions that can be used to compare a sample with a reference probability distribution or to compare two samples. To quantify the differences between these distributions, earth mover's distance (EMD) [22, 23] was calculated. EMD is a method for evaluating the dissimilarity between two probability distributions. Intuitively, given two distributions, one can be seen as a mass of earth properly spread in space and the other as a collection of holes in the same space. Then the EMD calculates the least amount of work needed to fill the holes with the earth. In this way, the value of EMD corresponding to the amount of earth multiplies the distance by which it moved. Therefore, the lower the EMD between two distributions, the higher the similarities between them.

6.10 Analysis of African Americans Admixture Dynamics

After filtering out the high-linkage SNPs using PLINK [9], we reduced the original SNPs to 341,672 SNPs. PCA was performed on all the samples at the individual level using these thinned markers. FRAPPE was performed on the 491, 557 autosomal SNPs successfully genotyped in all the samples of African American, CEU, and YRI. For the inference of ancestral chromosomal segments, haploid data of 88 YRI and 88 CEU from HapMap [6] were taken to represent their African and European parental population, respectively. The genetic contribution of Europeans to African Americans was set at 21.65 % at the population level based on FRAPPE results [19]. With the highest overall likelihood, the number of generations since admixture (λ) was taken as its estimation. By running HAPMIX in diploid model, we obtained the haplotypes and ancestral chromosomal segments of each African American individual.

The aforementioned forward-time simulation program was used to simulate admixture dynamics of African Americans. We considered the distribution of LACS for both ancestries simultaneously instead of only considering those of a single ancestry to infer the population admixture dynamics. For CGF model, the case in which the European population served as CGFD and the African population as CGFR was called CGF1 model, while the case in which the European population served as CGFR and the African as CGFD was called CGF2 model. The effective population sizes (N_e) of each population were set according to the HapMap [6]. In particular, the N_e of African, European, and African American populations were set at 17,094, 11,418, and 17,094, respectively. The contribution of European ancestry to African Americans was set at 21.65 % according to the observation of the 1,890 African Americans. Based on the recorded history of African Americans, the number of generations since admixture was set from

10 to 17, stepped by one generation for each model. For both empirical and simulated data, LACS < 0.5 cM for African ancestry and LACS < 0.8 cM for European ancestry were filtered out. The EMDs between the empirical distribution of LACS and that of each simulated dataset were calculated. Only the simulation showing the lowest EMD with empirical distributions for both ancestries was regarded as fitting the corresponding model.

6.11 Admixture Time and Admixture Proportion for Each Individual

Traditionally, the number of generations since admixture is a very useful concept in population genetics and is always calculated at the population level. Here, we treated each individual as a population with one sample and calculated the number of generations since admixture individual by individual. Based on the detected chromosomal segments of distinct ancestry in each individual, we were able to determine the number of ancestry transition events. Then we computed the number of generations since admixture for each individual (g) as follows [18]:

$$g = \frac{N}{4u(1-u) \times 35.4}$$

where N is the number of ancestry transitions observed and μ is the ancestry proportion of the admixed individual. A fraction of $2u(1-u)$ of all recombination events was expected, and 35.4 Morgan was the overall genetic distance used in our genetic map, which was calculated using HapMap genetic map.

The relationship between the number of generations and the admixture proportion for each individual is shown in Fig. 3.7. We found that individuals with extremely high amount of African or European ancestral components tended to have longer admixture time compared with the other individuals, which may indicate that individuals with a few European or African ancestral components might have inherited those components from much earlier admixture events, while those individuals with relatively even ancestral contributions were more likely to have descended from more recent admixture events.

6.12 Analysis of Mexican Admixture Dynamics

Two datasets were obtained: the 300 Mexican Mestizos and 239 individuals from their putative parental populations with autosomal 99,953 SNPs, and the 158 Mexican Americans with 767,454 autosomal SNPs. After a serial of filter, we had 36,000 autosomal SNPs shared by the remaining 652 samples (413 Mexicans and 239 samples from parental populations). PCA was performed at the individual level using all 36,000 autosomal SNPs shared by the 652 samples. We also performed STRUCTURE analysis on the 652 samples using SNPs with inter-marker distance >2 Mb. Taking advantage of the genome-wide high density SNPs data, we analyzed Mexican-Americans and Mexican-Mestizos separately for inference of ancestral chromosomal segments. For Mexican-Mestizos, we ran HAPMIX using haploid data of 30 Zapotecas (Amerindian) and 30 CEU as representatives of their Amerindian and European ancestry, respectively. For Mexican-Americans, haploid data of 44 AMI from HGDP and 44 CEU from HapMap were taken to represent Amerindian and European ancestry, respectively. The distribution of LACS of Mexicans was obtained by simply pooling together those of Mexican-Mestizos and Mexican-Americans.

The simulations performed on Mexicans were the same as those performed on African Americans, but some population parameters were changed to fit the known history of Mexicans. For the CGF model, the case in which the European population served as CGFD and the Amerindian as CGFR was named CGF1 model, while the case in which European population served as CGFR and Amerindian as CGFD was named CGF2 model. The N_e of the European, Mexican, and Amerindian populations was set at 11,418, 15,000, and 11,418, respectively. Based on the known Mexican history, the number of generations since admixture was set from 15 to 25, stepped by one generation for each model. The genetic contribution of European ancestry to Mexicans was set at 49.2 %, according to the empirical analysis of the 413 Mexicans. For both empirical and simulated data, LACS < 1.2 cM for either ancestry were filtered out. Finally, we also compared the distributions of LACS between Mexican-Mestizos and Mexican-Americans.

6.13 Simulation of African Americans and Its Parental Populations

Under selective neutrality, genetic drift of the admixed population and its parental populations contribute to the variation of ancestry proportion and population differentiation among loci [4, 17]. Therefore, we performed an extensive forward-time simulation to explore the potential impact of genetic drift on the locus-specific population differentiations between AAF and YRI, as well as the locus-specific population differentiation between AEU and CEU. The aforementioned phased data of 88 YRI and 88 CEU were taken, respectively, as the genotypes of common African ancestry and common European ancestry before population admixture. We simulated individuals of African Americans by constructing their genomes from a mosaic of haploid YRI and haploid CEU genomes. The generations since admixture (λ) were set to 14 according to CGF model based on previous reports, which was also supported by our observations. The gene flow (α) that African ancestry received from European each generation was calculated by $\alpha = 1 - (m_1)^{1/t}$, in which m_1 represents the mean proportion of European ancestry in African Americans. Both African and European parental populations evolved 14 generations simultaneously. Finally, genotypes of 113 Europeans, 113 Africans, and 1,890 African Americans, as well as their SNP-specific ancestral status, were output to match the sample sizes of the data. Based on the primary simulation, we performed extended simulations by setting a bottleneck event in the first generation during which N_e for African Americans were reduced to 8,000, 5,000, 3,000, 2,000, and 1,000, respectively.

6.14 Function Annotations and Ingenuity Pathway Analysis (IPA)

Genomic regions that deviated from genome-wide ancestral contributions or with extremely high F_{ST} were annotated based on HapMap website [27]. The SNPs showing substantial population differentiation were interrogated for network and functional interrelatedness using the Ingenuity Pathway Analysis (IPA) version 8.5 software tools. This software searches for information on genes in Ingenuity Pathways Knowledge Base, a repository of molecular interactions, regulatory events, gene-to-phenotype associations, and chemical knowledge, all collected from the full text of the peer-reviewed life sciences literature. With IPA, we can analyze data in the context of molecular mechanisms, identify key mechanistic differences between subpopulations, and further relate molecular events to higher-order cellular and disease processes.

6.15 Databases for Genes Underlying Monogenic and Complex Diseases

Two sets of human monogenic disease genes were investigated in this study. The first set of genes was obtained from the Online Mendelian Inheritance in Man database (OMIM; August 19, 2010 updated) [28, 29], which is the most comprehensive, authoritative, publicly available repository of human disease phenotypes and related genes. Overall 3,261 genes with disease-causing mutations confirmed in at least two laboratories or in multiple families were compiled in the database. The second dataset consisted of 968 genes in hand-curated OMIM database (referred to as hOMIM) from the study by Blekhman et al. [30], which is a highly confident monogenic disease database and has been extensively used in various studies recently [30–32].

Two sets of genes associated with human complex diseases were also investigated in this study. First, we obtained a list of genes associated with complex diseases from genetic association database (GAD, September 2, 2010 updated) [33, 34], which is an archive of human genetic association studies of complex diseases. We only included 3,117 'Positive Only' genes in GAD, of which 1,799 genes were also confirmed in the online catalog of published genome-wide association studies in NHGRI (GWAS catalog; June 8, 2010 update) [35], which contains the results of GWAS using at least 100,000 single nucleotide polymorphisms (SNPs).

In this study, we used the comprehensive collection of 21,526 reference genes from UCSC (hg18) as a representative set of all well-characterized human genes. To investigate the characteristics of genes underlying both monogenic and complex diseases, 524 genes in both hOMIM and GAD were considered as MC (monogenic and complex disease) genes; 444 hOMIM genes which are not in GAD were considered as monogenic (monogenic but not complex disease) genes; 2,593 GAD genes which are not in hOMIM were considered as Complex (complex but not monogenic disease) genes. Then all non-disease (ND) genes were considered as essential genes and OTHER genes. Although most studies used housekeeping genes based on gene expression data to approximate essential genes [36–39], we found that only 13 genes were consistent in four published studies. Therefore, it is not appropriate to use them in this study. Instead, we used a set of 1,520 genes for which knockouts were either inviable or caused sterility in mouse to represent human essential genes [40]. We classified those genes not belonging to any of the aforementioned functional gene categories as OTHER genes.

6.16 Protein–Protein Interaction Data

In this study, we used two protein–protein interaction databases, both of which were derived from the literature of high-quality experimental results and were manually curated by expert biologists. The first database is Human Protein Reference Database (HPRD) version 9 [41], which contains 9,673 proteins having annotation with at least one or more direct interacting proteins (39,240 non-redundant protein–protein interactions). It is currently the most comprehensive database of this kind and has been frequently used in various studies [37, 42]. The second protein–protein interaction database was derived from two highly qualified systematic yeast two-hybrid experiments (referred to as YTH) [43], containing 22,052 non-redundant interactions among 7,533 genes. HPRD and YTH can be viewed as being complementary to each other. In addition, we constructed a more comprehensive protein–protein interaction database by combining HPRD and YTH (referred to as HPRD-YTH). Finally, we constructed three networks based on HPRD, YTH, and HPRD-YTH, respectively. The gene connectivity of each gene in HPRD-YTH network is available.

6.17 Analysis of Gene Expression Pattern

The human expression dataset was downloaded from BioGPS database (http://biogps.gnf.org) [44], which is an extensible and customizable gene annotation portal. It includes basal expression signal intensity of 44,776 probe sets across 84 human tissues (known as Human U133A/GNF1H Gene Atlas). Probe sets shared by different genes were eliminated in order to avoid mistiness. Average gene expression signal intensity was used when a gene has more than one probe set. In this study, we measured three properties of gene expression pattern: expression level, highest expression signal ($S_{\max}$), and tissue specificity. Expression level is defined as the average signal intensity (S) of a human gene across all the 84 examined tissues. $S_{\max}$ is the highest expression signal of the gene across all tissues. And we used tissue specificity index τ [45] to measure the tissue specificity of a gene, which is defined as the heterogeneity of its expression level across all the tissues and is estimated by

$$\tau = \frac{\sum_{j=1}^{n}\left(1 - \frac{\log_2 S(j)}{\log_2 S_{\max}}\right)}{n-1}$$

where n (=84) is the number of tissues examined here. The values of τ range from 0 to 1, with higher values indicating higher variations in expression across tissues and thus higher tissue specificities. The final dataset contained 15,246 reference genes with expression level, $S_{\max}$, and tissue specificity for investigation.

6.18 Gene Ontology Annotation Analysis Using DAVID

Database for Annotation, Visualization, and Integrated Discovery (DAVID) [46, 47] provides a comprehensive set of functional annotation tools for investigators to understand the biological meanings behind large list of genes. To examine whether particular gene classes were enriched in our gene categories, we conducted enrichment analyses of essential genes, MC genes, complex genes, and monogenic genes using DAVID (version 6.7). The annotations for our four gene categories are available.

6.19 Structural Annotations and Evolutionary Rates of Human Genes

The structural annotations and pair-wise synonymous (d_S) and non-synonymous (d_N) distances between human and other species were obtained from Ensembl by BioMart [48, 49]. We firstly retrieved structural annotations such as chromosome positions, coding sequence (CDS) lengths, transcript number, 5′ UTR start, 5′ UTR end, 3′ UTR start, and 3′ UTR end. Then we retrieved pair-wise d_S and d_N distances between the orthologous genes of human–chimpanzee, human–orangutan, human–rhesus, human–mouse, human–cow, human–panda, human–pig, human–rabbit, and human–rat. The values of d_S and d_N from BioMart were calculated for coding sequence alignments by using the maximum likelihood method implemented in PAML [50]. To map genes on our compilations of gene categories, we used gene ID conversion tool in DAVID [46, 47] to transform Ensembl gene ID into official gene symbol.

First, the pair-wise orthologous genes with the highest identity were kept when there were multiple pair-wise orthologous genes for each human gene. Second, the pair-wise orthologous genes were removed when their d_S values were equal to zero. Finally, 17,044 pair-wise human–chimpanzee orthologous genes, 16,250 pair-wise human–orangutan orthologous genes, 16,424 pair-wise human–rhesus orthologous genes, 16,137 pair-wise human–mouse orthologous genes, 15,802 pair-wise human–cow orthologous genes, 16,084 pair-wise human–panda orthologous genes, 13,065 pair-wise human–pig orthologous genes, 14,865 pair-wise human–rabbit orthologous genes, and 15,576 pair-wise human–rat orthologous genes were obtained. The orthologous gene pairs with >80 % identities of the human gene length were kept for the evolutionary analyses. We also changed the identity threshold to 90 % and did similar analyses.

6.20 Human Polymorphism Data Analysis

HapMap data (Public Release #27) [6, 51] was used in population differentiation analysis. We removed the relatives in YRI, CEU, and CHB based on sample information and PLINK analysis (IBD > 0.2). Overall 113 unrelated YRI, 113 unrelated CEU, and 84 unrelated CHB were kept. We filtered out 3,836,272 autosomal SNPs genotyped in all three populations for analysis, and we also removed the monopolymorphic SNPs considering all samples. F_{ST} of each gene was calculated using the unbiased estimator proposed by Weir and Cockerham [17].

We calculated integrated haplotype score (iHS) [52] to detect recent positive selection in this study. The ancestral information of each SNP was obtained from UCSC [53]. The computer program calculating iHS was downloaded from the webpage of Pritchard's Laboratory (http://hgdp.uchicago.edu/Software), and phased data of 60 CEU were downloaded from HapMap (Public Release #22) [51]. We excluded data of X chromosome from our genome-wide analysis since population genetics of the X chromosome differs from that of the autosomes. The iHS value was calculated for each SNP with minor allele frequency (MAF) > 5 %, treating each SNP in turn as a core SNP. The proportion of SNPs in each gene with $|iHS| > 2$ was computed and thought of as iHS of the gene. Genes with ≤ 5 SNPs were filtered out in order to obtain reliable values.

In addition, we analyzed the re-sequencing polymorphic data of 8,427 genes in the Applera dataset [54], a genome-wide re-sequencing effort, considering European-American only. Tajima's D was estimated on the genes with ≥ 5 polymorphic SNPs in each population, respectively. Fu and Li's D^* was also calculated on the same data. We also analyzed Tajima's D of each gene from NIEHS SNPs, which were available from their official website.

6.21 Statistical Analyses

All statistical analyses and graphics were performed using R version 2.9 [25]. Kolmogorov–Smirnov tests were performed to compare the empirical distributions of locus-specific F_{ST} with those simulated. And chi-squared (χ^2 test) tests were performed to test the over-representation of each SNP category compared with non-genic SNPs among the high F_{ST} bin. To assess whether monogenic disease genes are more likely to be involved in complex diseases than non-monogenic disease genes, Fisher's exact test was used. To examine whether there are significantly differences among different gene categories regarding the values of associated diseases and phenotypes, protein connectivity, expression level, tissue specificity, evolutionary rate, F_{ST}, iHS and Tajima's D, Wilcoxon matched-pairs signed-rank tests were performed.

References

1. Chakraborty R, Weiss KM. Admixture as a tool for finding linked genes and detecting that difference from allelic association between loci. Proc Natl Acad Sci USA. 1988;85:9119–23.
2. Ewens WJ, Spielman RS. The transmission/disequilibrium test: history, subdivision, and admixture. Am J Hum Genet. 1995;57:455–64.
3. Pfaff CL, et al. Population structure in admixed populations: effect of admixture dynamics on the pattern of linkage disequilibrium. Am J Hum Genet. 2001;68:198–207.
4. Long JC. The genetic structure of admixed populations. Genetics. 1991;127:417–28.
5. Guo W, Fung WK. The admixture linkage disequilibrium and genetic linkage inference on the gradual admixture population. Yi Chuan Xue Bao. 2006;33:12–8.
6. Altshuler DM, et al. Integrating common and rare genetic variation in diverse human populations. Nature. 2010;467:52–8.
7. Wright S. Evolution in Mendelian populations. Genetics. 1931;16:97–159.
8. Li JZ, et al. Worldwide human relationships inferred from genome-wide patterns of variation. Science. 2008;319:1100–4.
9. Purcell S, et al. PLINK: a tool set for whole-genome association and population-based linkage analyses. Am J Hum Genet. 2007;81:559–75.
10. Bryc K, et al. Genome-wide patterns of population structure and admixture in West Africans and African Americans. Proc Natl Acad Sci USA. 2010;107:786–91.
11. Silva-Zolezzi I, et al. Analysis of genomic diversity in Mexican Mestizo populations to develop genomic medicine in Mexico. Proc Natl Acad Sci USA. 2009;106:8611–6.
12. Patterson N, Price AL, Reich D. Population structure and eigenanalysis. PLoS Genet. 2006;2:e190.
13. McVean G. A genealogical interpretation of principal components analysis. Plos Genetics. 2009;5.
14. Tang H, Peng J, Wang P, Risch NJ. Estimation of individual admixture: analytical and study design considerations. Genet Epidemiol. 2005;28:289–301.
15. Falush D, Stephens M, Pritchard JK. Inference of population structure using multilocus genotype data: linked loci and correlated allele frequencies. Genetics. 2003;164:1567–87.
16. Pritchard JK, Stephens M, Donnelly P. Inference of population structure using multilocus genotype data. Genetics. 2000;155:945–59.
17. Weir BS, Cockerham CC. Estimating F-statistics for the analysis of population structure. Evolution. 1984;38:1358–70.

18. Price AL, et al. Sensitive detection of chromosomal segments of distinct ancestry in admixed populations. PLoS Genet. 2009;5:e1000519.
19. Jin W, et al. Genome-wide detection of natural selection in African Americans pre- and post-admixture. Genome Res. 2012;22:519–27.
20. Scheet P, Stephens M. A fast and flexible statistical model for large-scale population genotype data: applications to inferring missing genotypes and haplotypic phase. Am J Hum Genet. 2006;78:629–44.
21. Rubner Y, Tomasi C, Guibas LJ. A metric for distributions with applications to image databases. 1998; 59–66 (IEEE).
22. Hitchcock FL. The distribution of a product from several sources to numerous localities. J Math Phys. 1941;20:224–30.
23. Rubner Y, Tomasi C, Guibas LJ. The earth mover's distance as a metric for image retrieval. Int J Comput Vis. 2000;40:99–121.
24. Levina E, Bickel P. The earth mover's distance is the Mallows distance: some insights from statistics. 2001;2:251–256 (IEEE).
25. Ihaka R, Gentleman R. R: a language for data analysis and graphics. J Comput Graph Stat. 1996;5:299–314.
26. Lilliefo Hw. On Kolmogorov-Smirnov test for normality with mean and variance unknown. J Am Stat Assoc. 1967. 62:399–& .
27. Thorisson GA, Smith AV, Krishnan L, Stein LD. The international HapMap project web site. Genome Res. 2005;15:1592.
28. McKusick VA. Mendelian inheritance in man and its online version, OMIM. Am J Hum Genet. 2007;80:588–604.
29. Amberger J, Bocchini CA, Scott AF, Hamosh A. McKusick's online mendelian inheritance in man (OMIM). Nucleic Acids Res. 2009;37:D793–6.
30. Blekhman R, et al. Natural selection on genes that underlie human disease susceptibility. Curr Biol. 2008;18:883–9.
31. Cai JJ, Borenstein E, Chen R, Petrov DA. Similarly strong purifying selection acts on human disease genes of all evolutionary ages. Genome Biol Evol. 2009;2009:131–44.
32. Huang N, Lee I, Marcotte EM, Hurles ME. Characterising and predicting haploinsufficiency in the human genome. PLoS Genet. 2010;6:e1001154.
33. Becker KG, Barnes KC, Bright TJ, Wang SA. The genetic association database. Nat Genet. 2004;36:431–2.
34. Zhang Y, et al. Systematic analysis, comparison, and integration of disease based human genetic association data and mouse genetic phenotypic information. BMC Med Genomics. 2010;3:1.
35. Hindorff LA, et al. Potential etiologic and functional implications of genome-wide association loci for human diseases and traits. Proc Natl Acad Sci USA. 2009;106:9362–7.
36. Tu Z, et al. Further understanding human disease genes by comparing with housekeeping genes and other genes. BMC Genom. 2006;7:31.
37. Podder S, Ghosh TC. Exploring the differences in evolutionary rates between monogenic and polygenic disease genes in human. Mol Biol Evol. 2010;27:934–41.
38. Eisenberg E, Levanon EY. Human housekeeping genes are compact. Trends Genet. 2003;19:362–5.
39. Haverty PM, et al. HugeIndex: a database with visualization tools for high-density ligonucleotide array data from normal human tissues. Nucleic Acids Res. 2002;30:214–7.
40. Liao BY, Scott NM, Zhang J. Impacts of gene essentiality, expression pattern, and gene compactness on the evolutionary rate of mammalian proteins. Mol Biol Evol. 2006;23:2072–80.
41. Keshava Prasad TS, et al. Human protein reference database–2009 update. Nucleic Acids Res. 2009;37:D767–72.
42. Kandasamy K, et al. Human Proteinpedia: a unified discovery resource for proteomics research. Nucleic Acids Res. 2009;37:D773–81.
43. Goh KI, et al. The human disease network. Proc Natl Acad Sci USA. 2007;104:8685–90.

44. Wu C, et al. BioGPS: an extensible and customizable portal for querying and organizing gene annotation resources. Genome Biol. 2009;10:R130.
45. Yanai I, et al. Genome-wide midrange transcription profiles reveal expression level relationships in human tissue specification. Bioinformatics. 2005;21:650–9.
46. Dennis G Jr, et al. DAVID: database for annotation, visualization, and integrated discovery. Genome Biol. 2003;4:P3.
47. da Huang W, Sherman BT, Lempicki RA. Systematic and integrative analysis of large gene lists using DAVID bioinformatics resources. Nat Protoc. 2009;4:44–57.
48. Hubbard TJ, et al. Ensembl 2009. Nucleic Acids Res. 2009;37:D690–7.
49. Haider S, et al. BioMart central portal–unified access to biological data. Nucleic Acids Res. 2009;37:W23–7.
50. Yang Z. PAML: a program package for phylogenetic analysis by maximum likelihood. Comput Appl Biosci. 1997;13:555–6.
51. Frazer KA, et al. A second generation human haplotype map of over 3.1 million SNPs. Nature. 2007;449:851–61.
52. Voight BF, Kudaravalli S, Wen XQ, Pritchard JK. A map of recent positive selection in the human genome. PLoS Biol. 2006;4:446–58.
53. Rhead B, et al. The UCSC Genome Browser database: update 2010. Nucleic Acids Res. 2010;38:D613–9.
54. Bustamante CD, et al. Natural selection on protein-coding genes in the human genome. Nature. 2005;437:1153–7.

Web Resource

The URLs for data and software presented herein are as follows:

Coriell Cell Repositories, http://ccr.coriell.org
The International HapMap Project, http://www.hapmap.org/
The HGDP-CEPH project, http://www.cephb.fr/en/hgdp
Mexican Genetic Diversity Project (MGDP), ftp://ftp.inmegen.gob.mx/
Illumina iControl Database (iControlDB), http://www.illumina.com/science
PLINK v1.06, http://pngu.mgh.harvard.edu/purcell/plink/
EIGENSOFT, http://genepath.med.harvard.edu/~reich/Software.htm
FRAPPE and SABER, http://med.stanford.edu/tanglab/software/
fastPHASE, http://stephenslab.uchicago.edu/software.html
PHASE program, http://stephenslab.uchicago.edu/software.html
STRUCTURE, http://pritch.bsd.uchicago.edu/software.html
HAPMIX, http://www.stats.ox.ac.uk/~myers/software.html
Annotation of SNPs, http://genome.ucsc.edu/
HAPMIX, http://www.stats.ox.ac.uk/~myers/software.html
HAPAA, http://hapaa.stanford.edu/HAPAA%20Project.html
LAMP and LAMP-ANC, http://lamp.icsi.berkeley.edu/lamp/
Ensembl Biomart, http://www.ensembl.org/biomart
OMIM genes, http://www.nslij-genetics.org/search_omim.html
BioGPS, http://biogps.gnf.org
The Genetic Association Database (GAD), http://geneticassociationdb.nih.gov/
Human Protein Reference Database (HPRD), http://www.hprd.org/
NIEHS SNPs, http://egp.gs.washington.edu/
GWAS catalog, http://www.genome.gov/admin/gwascatalog.txt
OMIM genes, http://www.nslij-genetics.org/search_omim.html
R, http://www.r-project.org/

W. Jin, *Admixture Dynamics, Natural Selection, and Diseases in Admixed Populations*, Springer Theses, DOI 10.1007/978-94-017-7408-6

Epilogue

Firstly, I would like to say that life in the late of my Ph.D. program was really tough. In those days, the pressure in trying to complete the Ph.D. dissertation nearly drove me crazy, and the life is boring only with my computer stay nearby. Looking back now, I was really surprised that I had gone through. Luckily, the results turned out to be good after all these hard work. I am proud that three or four of my works can be published on high reputation journals.

What I have presented in this thesis is just the tip of the iceberg. In fact, I have explored tens of research projects in the past few years, although most of which have ended up unfinished or unpublished. I do not regret them because I have learned much in the process of doing research. Although many of them might be negligible or too difficult to complete, the ideas underlying each of them used to be thought as great and glorious. I really miss the life at the beginning of my Ph.D. project when there is little pressure. I could still feel the excitement when I started my first project in Felix's lab, under the guidance of Dr. Xu, and the frustration when the programming problem obstructed the way of my project, and the happiness when I completed the project at the end of my first rotation.

In fact, my style in doing research might be inherited from my supervisors Felix and Dr. Xu. Felix always made science fun for me, with his enthusiasm and inspiration. I still remember many interesting stories about his experiences in research and science, which are enlightening, especially the one he told in his office on Liren Biology Building when I did my first rotation. I really appreciate Felix for his ideas, patience, kindness, and academic experience, which are invaluable to me and may benefit me in my whole scientific life. I still remember the first meeting with Dr. Xu in which he told me that I could join with him only if I was interested in his work. Room 312, where he met me and used to be his office, is the place where I devoted most of my time. I also thank him for his guidance and supervision. I am also thankful for the excellent examples both of the two supervisors have set themselves as successful population geneticists. Finally, I really appreciate their patience for revising my manuscripts, which could not be published without their hard work.

W. Jin, *Admixture Dynamics, Natural Selection, and Diseases in Admixed Populations*, Springer Theses, DOI 10.1007/978-94-017-7408-6

I have been in Shanghai for almost 6 years since I did my undergraduate dissertation in the Institute of Biochemistry and cell Cell Biology, SIBS, CAS. I really favor this beautiful dynamic city with various cultures and people, during which time I experienced many novel things. Firstly, I appreciated Academician Enduo Wang in Institute of Biochemistry and Cell Biology, who provided the opportunity and basic living conditions for me being an intern in her lab. I also thank Prof. Mofang Liu for her recommendation and her guidance. I might not come and stay in Shanghai without their help. The 6 years in Shanghai is the most important period in my life, during which I experienced a fateful change from an ignorant college student to a young scientist, made a lot of friends who that had similar dreams, attended many conferences, met lots of famous scientists, experienced my first real love, and had a life direction. I think there will never be a second chance for me to get so many things in such a short period.

I also really enjoy the academic environment in SIBS, which may be the best in the world according to my limited experiences. The campus is always very quiet and peaceful even in the central area of the modern and dynamic city. I still remember the good days I walked with Jing around the campus. Since there may be still a few months before I leave, maybe I can fully enjoy a small period of wonderful life in the following days after the thesis defense. I think I will miss everything here no matter where I will be in the future.

Wenfei Jin
Nov 2, 2011
SIBS main building, 320 Yueyang Rd, Shanghai

MIX
Papier aus verantwortungsvollen Quellen
Paper from responsible sources
FSC® C105338

If you have any concerns about our products,
you can contact us on
ProductSafety@springernature.com

In case Publisher is established outside the EU,
the EU authorized representative is:
Springer Nature Customer Service Center GmbH
Europaplatz 3, 69115 Heidelberg, Germany

Printed by Libri Plureos GmbH
in Hamburg, Germany